NATO's Feuding Members: The Cases of Greece and Turkey

INTERNATIONALE SICHERHEIT

Herausgegeben von Heinz Gärtner

Band 5

PETER LANG

Frankfurt am Main · Berlin · Bern · Bruxelles · New York · Oxford · Wien

Hakan Akbulut

NATO's Feuding Members:
The Cases of Greece
and Turkey

PETER LANG
Europäischer Verlag der Wissenschaften

Bibliographic Information published by Die Deutsche Bibliothek
Die Deutsche Bibliothek lists this publication in the Deutsche Nationalbibliografie; detailed bibliographic data is available in the internet at <http://dnb.ddb.de>.

Gedruckt mit Förderung des Bundesministeriums
für Bildung, Wissenschaft und Kultur in Wien.

ISSN 1618-5706
ISSBN 3-631-53539-2
US-ISBN 0-8204-7683-8

© Peter Lang GmbH
Europäischer Verlag der Wissenschaften
Frankfurt am Main 2005
All rights reserved.

Printed in Germany 1 2 3 4   6 7

www.peterlang.de

ACKNOWLEDGEMENTS

First, I would like to thank Prof. Heinz Gärtner for his assistance and for making this publication possible. This publication would also not have been possible without the funding of the Federal Ministry for Education, Science and Culture and the efforts of the Peter Lang Publishing Group.

I am also very much obliged to Greek, Turkish, and NATO officials who took their precious time to talk to me and have been very helpful. I would also like to thank those people who helped me to arrange these appointments.

I am also deeply indebted to Fatma and our families and friends; to Ersin Akbulut, to Christina Sue Augsburger and David.

dedicated

to the memory of my father

&

to my mother

CONTENTS

1) INTRODUCTION

In the wake of the European Union's (EU) December 2004 summit, the public in Europe has been witnessing a furious debate on the question as to whether the Union should start accession negotiations with Turkey, a country that has been pursuing the goal of membership for more than four decades. While the outgoing EU Commission presented its recommendation on the issue in early October 2004 calling for the opening of accession talks and the heads of government and state of the member-states are likely to follow this recommendation on December 17th and decide on a date for the beginning of talks, the issue of Turkish membership is likely to continue to divide Europeans. Even though the Commission offered the view that Turkey had fulfilled the Copenhagen criteria and delivered a report in favor of accession talks, it nevertheless made clear that the process would be "open-ended" and that Turkey's membership could not be taken for granted at the end of this negotiation process. Simultaneously, governments and political parties across Europe have still not been able to iron out their differences related to the "Turkish question". There are still those like the British PM, Tony Blair, or the German Chancellor, Gerhard Schröder, who support Turkey's EU bid and others - such as Andrea Merkel of the German Christian Democratic Union - opposing the idea and propagating the idea of a "privileged partnership" or of a "loose association" instead. Thus the debate on Turkish membership is likely to continue even after December 17th, and we are bound to repeatedly hear the same arguments in favor of or against a Turkish membership.

The debate so far has concentrated on the *different character* of Turkey as a mainly Muslim country, on the consequences of membership on the decision-making procedures in European institutions, on the amount of subsidies Turkey would receive from the EU or the number of Turks that would head towards the older EU states once the country has joined the Union, as well as on the contribution Turkey could make to the foreign and security policy of a Union which is willing to become a greater global player. Overall, those opposing membership have usually highlighted the budgetary consequences of eventual Turkish membership and the problems member-countries have in integrating Turkish citizens already in the country, while those favoring Turkish membership have pointed out that Europe needs to prove that it is not an exclusive "Christian club" and that Turkey, given its strategic importance and military power, could make major contributions to Europe's nascent defense and security policy.[1]

---

[1] Arguments in favor and against Turkish membership can be found in the paper prepared by the EU Commission titled "Issues Arising from Turkey's Membership Perspective" (http://www.europa.eu.int/comm/enlargement/report_2004/pdf/issues_paper_en.pdf) or in the

An issue that has been dealt with only superficially or not at all during this debate, but which now will apparently play a decisive role on December 17th and in its aftermath, is the dispute between Turkey and the EU members Greece and Cyprus. Despite the fact that there has been little to no reference to this relationship in the public debate when elaborating on the issues arising from Turkey's membership perspective, according to press reports, the draft document prepared by the Dutch Presidency, which will constitute the basis of debate on December 17th, asks Turkey to recognize the Greek Cypriot state, at least tacitly, if it wants the talks to commence (cf. IHT, Radikal, Hürriyet, TDN, November 30th, 2004). Turkey will be required to amend the 1963 Ankara Agreement of association with the European Economic Community (EEC) in a manner taking account of the 10 new members that joined the Union in May 2004. Likewise, Turkey will be asked to commit itself to "good neighborly relations" and to the "resolution of remaining border disputes" as a precondition. In the view of EU diplomats, these requirements amount to a recognition of the Greek Cypriot administration (IHT, TDN, November 30th, 2004). The Turkish Foreign Minister, Abdullah Gül, on the other hand, dismissed the leaking of the document to the press as mere tactics and asserted that Turkey would accept no additional conditions. With regard to the amending of the association agreement in order to include Cyprus, he added that the issue could be paid attention only after the negotiations had started and Turkey's expectations had been met. Two days later, he openly stated that Turkey would not recognize the Greek Cypriot state prior to a settlement (Hürriyet, Radikal, December 2nd, 2004).

In fact, the prospect of Turkish membership and Greek support for the EU bid of the neighbor to the east have contributed to an improvement in bilateral relations in recent years and have given impetus to Turkish efforts to solve the Cyprus issue. It is this giving rise to the general expectation that the prospect of or eventual Turkish membership will facilitate the finding of a comprehensive solution. In the same manner, the Independent Commission on Turkey which presented its report in favor of Turkish membership in September 2004, for example, offered the opinion that "the opening of accession talks with Turkey would greatly facilitate the search for solutions here (2004, 21)". With regard to Cyprus, the view was offered that, "[w]hatever turn developments in Cyprus may take in the coming years, it could confidently be predicted that at the latest, Turkey's accession to the EU would see the island's division brought to an end" (Ibid., 21). Even though the optimistic forecasts offered might be well-founded given the positive developments of recent years in the Greece-Cyprus-Turkey triangle, it cannot be taken for granted that Greek-Turkish relations will not

---

report of the Independent Commission on Turkey, "Turkey in Europe: More than a Promise?" (http://www.soros.org/resources/articles_publications/publications/turkey_2004901).

burden cooperation within EU's institutions or that eventual membership will have solely positive effects on the disputes between the countries in question and the two communities on Cyprus. Thus the consequences arising out of Greek-Turkish-Cypriot problems might constitute an important aspect that has been neglected in the debate so far, and it might not be unwise to pay greater attention to the issue.

In contrast to the current debate, questions arising from the membership of feuding countries in the same institution and its impact on bilateral relations had been of central importance during the debate on enlargement of the North Atlantic Treaty Organization (NATO) and the cases of Greece and Turkey were often cited. With the Soviet Union gone and the former communist bloc countries left on their own, many analysts had identified a security vacuum in the region that needed mending. There had been a general sense that Eastern Europe constituted a volatile area prone to conflict and that security and stability had to be projected into the region to the east of the former Iron Curtain. Officials in Western capitals as well as their counterparts in Eastern Europe had embraced the idea very early that institutions like NATO or the EU had to be extended eastwards to include former Warsaw Pact states. There existed a common understanding that these institutions had helped to secure peace and stability in Western Europe throughout the Cold War era and should now be enlarged to do the same for Eastern Europe.

While the proponents of NATO enlargement based their arguments on the assumption that NATO had not only deterred a Soviet attack but had simultaneously contributed to peaceful relations among its members, this wisdom was not unanimously embraced by scholars of international relations. In fact, theorists had even voiced doubts whether NATO would still exist in ten years time. Realists offered the view that it had not been NATO that had secured peace for the past 40 years but the balance of power between the two blocs in connection with nuclear deterrence. Moreover, with the Soviet threat gone, NATO had lost its *reason d'être*. Kenneth Waltz, probably the most outstanding representative of the neorealist school, concluded that NATO's years were numbered (1993, 76). The institutionalists, on the other hand, argued that NATO as an international institution providing a forum for consultations and guaranteeing a certain degree of transparency had contributed to the maintenance of peace and stability during the Cold War years. In their view, the conditions since the end of the Cold War had not worked against the existence of security institutions, but to the contrary necessitated their continued existence. However, they also did not take the continued existence of NATO for granted and asserted that NATO had to adapt itself to the changing circumstances and adopt new tasks if it wanted to escape the fate of the Warsaw Pact (see Hellmann and Wolf 1993, 21).

Indeed, NATO has not only survived, but has extended membership to 10 former east bloc countries, waged its first war ever in the Balkans, went out-of-area, even out of Europe, and invoked Article V for the first time in its history after the September 11th attacks on the United States. Nevertheless, the debate on the relevance of NATO and its role in inter-member relations has not ended, and scholars from both schools have continued to provide differing assessments.

The very same question over the impact of NATO membership on inter-member relations constitutes the object of research in this book, while the debate is limited to the cases of Greece and Turkey. Both countries have not been able to put an end to their disputes related to the Aegean Sea and the island of Cyprus and have found themselves on the brink of war on several occasions over the last 50 years irrespective of the fact that both have been members to NATO since 1952. Thus, their conflict represents a case where the differing expectations and assessment of scholars of international relations, bureaucrats and politicians alike can be put to the test. This book therefore pays attention to the theoretical debate, portraying the arguments of neorealists and neoliberal institutionalists who have been dominating this debate, and explores the question whether NATO helped to mitigate the conflict or, to the contrary, even contributed to the tensions in practice. Another point of interest will be the effect of NATO membership on the democratic development of these countries given the fact that this question constituted another dimension to the debate over NATO enlargement.

In the first part of this book, the reader will be presented with a synopsis of Greek-Turkish relations. Then, the main issues of the dispute will be summarized. Next, neorealist and neoliberal institutionalist views on cooperation and the role played by international institutions will be portrayed. This discussion of differing theories is expected to enable us to identify the points of deviation and agreement in these theories, to answer the question whether and how international institutions can promote peace and stability among member-countries, and also to enlist a number of *institutional factors* which can be employed to assess NATO's role in this conflict. Furthermore, the theoretical question as to whether institutions can facilitate democratization processes in member-countries will also be tackled.

Subsequently, the focus will shift to the Atlantic Alliance and NATO mechanisms for dealing with internal disputes will be portrayed, while assessing the role of the secretary-general. A presentation of NATO's record of involvement and mediation in the Greek-Turkish dispute will follow. In the following section, the effects that factors like issue-linkage, transparency, consultation, and military assistance have had on the course of relations between Greece and Turkey will be examined, and simultaneously close attention will be paid to the concept of *institutional capture* put forward by Ronald Krebs (1999). Moreover, the reader will be provided with an appraisal of the effects of NATO membership on the democratic development of both countries. The final section will provide a summary of the main findings of this book.

2) THE GRECO-TURKISH CONFLICT

2.1) Greek-Turkish Relations: A Synopsis

The animosity between Greeks and Turks can be traced back to the 11th century when the Turks entered Asia Minor and started to expand their empire at the cost of the Byzantine Empire. In 1453, the Ottomans captured Constantinople, the heart of the Byzantine Empire, and continued their seizure of the Balkans, eventually bringing the Hellenic mainland and the islands in the Aegean under their rule. The rule of the Ottomans was to last until 1831 when Greece gained its independence following a war of independence that had begun in 1821. Great Britain, France, and Russia had intervened in the conflict on behalf of the Greeks and forced the ailing Ottoman Empire to accept an agreement settling the new borders (cf. Clogg 2002, 7-45; Riemer 2000, 45-46).

From this point onwards, Greece started adding areas formerly under Ottoman rule to its territory. The *Megali Idea* (Great Idea), envisioning a Hellas encompassing all areas inhabited by Greeks and having its capital in Constantinople, constituted the main ideology of the newly established state and acted as the driving force behind Greek expansionism (see Clogg 2002, 47). The largest gains of territory were made during the Balkan Wars of 1912 and 1913 when Greek territories expanded by 70 percent and the total population increased from 2,800,000 to 4,800,000 (Ibid., 82). World War I then brought further territorial gains for Greece. After the War had ended, Greece landed troops on the Aegean coast of Turkey and captured the city of İzmir. Greek forces tried to push forward into areas further eastwards but were met with resistance by the Turkish forces. In September 1921, the Turks started a counter-offensive and forced the Greek forces to retreat and evacuate Asia Minor (Ibid., 96-97). This *catastrophe of Asia Minor* marked, more or less, the end of the *Megali Idea*. However, Greece was to make further territorial gains as late as 1947 when Italy agreed to cede the Dodecanese Islands to Greece.

The Lausanne Treaty of 1923 established the new frontiers between Greece and Turkey. Apart from territorial readjustments, both countries agreed on the exchange of populations forcing 1,110,000 Greeks living in Turkey and 380,000 Turks living in Greece to leave their homes and resettle in their respective motherlands (Ibid., 99-100). Both countries now aimed at reconciliation, and their foreign policy objectives seemed to be compatible. Their primary goal in the late 1920s and in the 1930s was the maintenance of the status quo as well as of their territorial integrity (Varvaroussis 1979, 25). In 1930, they signed the Ankara Convention and an additional protocol which aimed to prevent an arms race in the

Aegean (see Ibid., 27; Clogg 2002, 107; Stearns 1992, 89). In the face of Bulgarian and Italian[2] revisionism, Greece and Turkey, together with Yugoslavia and Romania, joined the Balkan Pact of 1934, guaranteeing the existing boundaries and obliging the members to consult each other when threats to regional peace emerged (Clogg 2002, 107; Kuniholm 1980, 16). Later, the Greek Prime Minister, Venizelos, even visited Turkey and nominated the Turkish President, Mustafa Kemal Atatürk, for the Nobel Peace Prize (Clogg 2002, 107). In the late 1930s, both countries signed another agreement obliging both parties to assist the other signatory when it faced aggression and to contribute to the peaceful resolution of the conflict (Fırat 2001a, 355). The fact that Greece could extend its territorial waters and its airspace from 3 to 6 miles and from 3 to 10 miles respectively in this period without giving rise to any protests in Turkey was illustrative of the atmosphere of friendship. The Turkish Foreign Minister, Aras, even stated during a speech in parliament that the Montreux Convention had lifted the restrictions on the militarization of Lemnos and Samothrake and that he welcomed this development (see Fırat 2001a, 356) - from the 1960s onwards, the militarization of these islands turned into one of the thorniest issues causing considerable tensions between the two countries.

The Second World War was to bring occupation by three different powers - namely by Germany, Bulgaria, and Italy - and devastation to Greece while Turkey succeeded in resisting pressures for joining the war and only declared war on Nazi Germany shortly before the war ended (Clogg 2002, 121). In Greece, occupation by foreign forces ended in October 1944, but this time a civil war broke out between the government and the communists who had played an important role in the resistance against the occupation forces (Ibid., 122-130). The Democratic Army of the communists initially received support from Yugoslavia, Bulgaria and Albania (Ibid., 139-141). However, after its exclusion from the Cominform, Yugoslavia withdrew its support for the communists and closed its border, while the Soviet Union, in an attempt not to burden relations with the United States and Great Britain, preferred a strategy of non-intervention. In the meantime, Greek regular forces received huge amounts of aid from the United States. In the end, the communists were defeated and the civil war ended in 1949.

The era of détente in Greek-Turkish relations, which had begun almost two decades before, was to last until the mid-1950s, in spite of the fact that Greece was blaming Turkey for having let down its neighbor during the war. Both

---

[2] In fact, initially, Italy had tried to contribute to a Greek-Turkish conciliation and bring these countries closer to itself in order to create a balance against Yugoslavia in the Balkans (Ülman 1968, 250).

countries had become recipients of US aid under the Truman Doctrine[3] formulated in 1947, both sent troops to Korea in 1950, and both joined NATO in 1952.[4] In the face of the common Soviet threat, Greece and Turkey, together with Yugoslavia, signed an agreement of friendship and cooperation in 1953, which then became a formal alliance in 1954 (Varvaroussis 1979, 31; Papacosma 2001, 200). This alliance was to be of short duration given the rapprochement between Yugoslavia and the Soviet Union and upcoming tensions between Greece and Turkey over the status of the Mediterranean island of Cyprus (see also Fırat 2001b, 591-592).

The relations between the two countries were to deteriorate from the mid-1950s onwards. It was the campaign for enosis that revived the hostility between them. Later, other contentious issues in the Aegean, such as the delimitation of the continental shelf, the breadth of Greek territorial waters and air space, the militarization of certain Greek islands, the sharing of command and control rights in NATO, and the treatment of the respective minorities added to the tensions between the two countries which have found themselves on the brink of war many times throughout the last decades. Of course, there were also short periods of détente like that experienced in the aftermath of the 1987 crisis and referred to as the *Davos Process* - the two prime ministers had met on the sidelines of a World Economic Forum meeting in Davos in 1988. At that time, both countries declared

---

[3] After Great Britain - which had signed a Declaration of Mutual Assistance with both countries in 1939 (Kuniholm 1980, 19) - declared it was no longer in a position to provide aid to Greece and Turkey, the US took over the job, and President Truman pledged to assist Greece and Turkey in their fight against communism in an address to the Congress. He asked Congress to release $ 400 million in aid to both countries and to send US civilian and military personnel into the region to supervise the use of this aid and provide help in reconstruction (Ireland 1981, 24-29).
Given the communist insurgency in Greece and Soviet revisionist aims on Turkey, the US feared these countries might be lost to the Soviets. In fact, Soviet troop movements in Bulgaria and across the Turkish-Iranian border had led the Americans to conclude that a Soviet invasion of Turkey was impending (Kuniholm 1980, 317, 359). In such a case, the whole region might have been lost to the Soviets. This, in turn, would have posed a threat to the flow of oil from the Middle East to Europe while simultaneously encouraging the position of communist parties in France and Italy. A communist takeover in France and Italy could, in turn, have entailed the loss of the whole of Western Europe to the communists.
[4] The Turkish government viewed the sending of troops to Korea as an inevitable measure to gain membership in NATO (Erhan 2001a, 545-546). Only a week after the government had decided to send troops to Korea, it was to reiterate its application for NATO membership, which was rejected again.
Turan holds that there can be no doubt that Turkey sent troops to Korea in order to show the US that it was a reliable ally, ready to make military contributions (2001, 652). This view is supported by Veremis, who asserts that both Greece and Turkey sent troops to Korea to override the objections raised by Great Britain and Scandinavian countries against their membership (1988b, 242).

their willingness to improve relations and agreed to establish committees that would explore the possibilities for cooperation in areas such as economy, trade and tourism, work out an inventory of contentious issues, and propose solutions to these problems. Furthermore, they made mutual gestures of good-will in the following months: Turkey declared it would annul a 1964 decree restricting the property rights of the Greek minority and lifted the visa obligation for Greek nationals, while Greece agreed to sign the adaptation protocol to the association agreement between Turkey and the European Community (Axt & Kramer 1990, 53-54; Clogg 1991, 21). Nevertheless, they failed to make progress on substantive issues, and differences – among others those related to NATO infrastructure funding - beset their relations, again putting an end to the short-lived period of détente.

The end of the Cold War was to expand the Greek-Turkish competition into new areas and to add to their mutual anxieties (see Riemer 2000, 49-51). With the rigid system of the Cold War gone, both countries could now try to play a greater role in the Balkans and Central Asia. Given their centuries old animosity, both countries were anxious- while this was truer for Greece- that the new setting could favor the opponent more and add to its capabilities and strategic weight, in turn putting it in a better position in their bilateral disputes. Moreover, both were concerned that the adversary could establish new alliances to further their own position within the framework of their dispute. In particular, the facts that Turkey took responsibility for the welfare of the Muslims in the Balkans, was quick to recognize Macedonia under its original name, "Republic Macedonia", which was contested by Greece, and enjoyed heartily relations with Albania, which Greece accused of mistreating the Greek minority, exacerbated Greek fears of being surrounded by an Islamic arc (see Kramer 2000, 149; Clogg 2002, 202-203, 208; Stivachtis 2000, 473).[5] Greece was concerned Turkey might form an anti-Greek coalition with these countries in dispute with Greece (Axt 1993, 1-2).[6] Moreover, Turkish policies aiming at establishing a special relationship with the Turkic republics of Central Asia[7] and recommending itself as the main route for the

---

[5] Thanos Veremis, the director of the Greek think-tank "Foundation for Defense and Foreign Policy" is said to have stated that Turkey was planning to annex Western Thrace in the long run (quoted in Vamik & Itzkowitz 1994, 166).
[6] Turkey signed a military cooperation agreement with Albania in 1992 (Uzgel 2001a, 55). In 1995, a military cooperation agreement was signed with Macedonia (Ibid., 60).
[7] Turkey concluded various agreements in the field of economy, culture, education, communication, technology and transport with the countries in the region (Kramer 2000, 92, 111). In addition, military agreements were signed with Azerbaijan and Georgia (Ibid., 106). Moreover, Turkey was viewed as a model for these newly independent countries. In the initial era, there was a common understanding in the Western world that the Turkish model should be promoted in the region as a counterweight to religious fundamentalism and Russian and Iranian dominance.

transportation of Caspian oil and gas to European markets as well as its strategic partnership with Israel acerbated Greece's perception of threat. In the meantime, Greece furthered its ties with Yugoslavia, sought military cooperation with Armenia, Syria,[8] Iran and Iraq, and concurred with Armenia and Iran to cooperate in the field of energy. It may come as no surprise that these countries Greece had chosen to cooperate with are viewed by Turkey as threats to its security. Turkey has sided with Azerbaijan in its dispute with Armenia concerning Nagarno Karabakh and almost waged a war against Syria, because the latter had been providing shelter to state enemy number one, Abdullah Öcalan. Moreover, Turkey has accused Iran of trying to export its Islamic revolution to Turkey and has been involved in a water dispute with Iraq and Syria. With regard to the new big issue of energy transportation, Greece and Turkey favored different routes for the oil pipelines planned to transport the oil of the Caspian region to European markets. While Turkey favored the construction of a pipeline along the Baku-Ceyhan route, Greece lobbied for the transportation of oil from the Russian port of Novorossiisk through Bulgaria to the Greek port of Alexandroupolis (cf. Bahçeli 2000, 465).

Nonetheless, contrary to fears that Greece and Turkey might be involved in open conflict in the Balkans given their opposed allegiances, both countries refrained from adventurous and provocative endeavors in the region. In fact, the crisis in the Balkans brought these countries closer to each other, as both had a stake in maintaining order and stability in the region. The fact that Greece, whose Balkan policies had initially been viewed as "obstructionist" in NATO and EU circles, began to adopt more pragmatic policies from 1995 onward - in part as a reaction to the growing Turkish influence in the region (Papacosma 2001, 210) - contributed to a changing atmosphere. Both preferred multilateral action to adventurism. Greece participated in the embargo imposed on former Yugoslavia and opened its bases for use during the NATO operation "Deny Flight". After the Dayton Agreement had been signed, both Greece and Turkey participated in I/SFOR "and displayed some cooperation in their assigned missions" (Papacosma 2001, 210). They participated in Operation Alba in an attempt to restore order in Albania. When NATO waged a war against Serbia, Turkey participated actively while Greece allowed the transfer of NATO troops to Macedonia through its territory (Papacosma 2001, 217). Both countries cooperated in the humanitarian

---

Turkish President Özal had declared that the 21st century would be the century of the Turks (Gumpel 2000, 21). Nevertheless, despite the contacts established with theses countries, Turkey was soon to be disillusioned. It lacked the assets to establish its hegemony in the region and had to recognize that it faced a mighty competitor, namely Russia. Moreover, the Turkic republics, while seeking closer ties with Turkey, were not interested in a paternalistic relationship.

[8] Syria and Greece were reported to have agreed that Greek planes could use Syrian airports in cases of emergency implying a Turkish attack against Cyprus (Bahçeli 2000, 465).

field during the crisis and have been participating in KFOR. Furthermore, they were represented in the NATO force in Macedonia. Greece and Turkey have also been cooperating within the framework of the Black Sea Economic Cooperation (BSEC)[9] and have, together with countries like Bulgaria and Romania, agreed on the creation of a Balkan peace-keeping force (Southeast European Multinational Force), which might carry out operations in the Balkans and elsewhere under the authority of the UN or the OSCE (Kramer 2000, 158-161; TDN, March 20th, 1998).[10]

Their multilateral approach and concurring views on the Balkans did not mean that Greeks and Turks had managed to sort out their differences on contentious issues. On the contrary, their relations remained highly tense until the second half of 1999. Aegean issues and developments on the island of Cyprus continued to burden relations. For example, in January 1996, both countries nearly went to war against each other over the status of a little islet in the Aegean. Yet, relations took another step forward in June 1997, when Greece and Turkey signed the Madrid Declaration under pressure from the US on the sidelines of a NATO meeting (Clogg 2002, 228-229). They declared willingness to respect the other's rights in the Aegean and to abstain from threatening the use of force (see also Papacosma 2001, 213). The Madrid Declaration was generally referred to as a "non-aggression pact" (see for example McDonald 2001, 137). However, the period of tranquility in relations was of short duration. Only a couple of months later, Greece claimed that armed Turkish fighter jets had intercepted a plane carrying the Greek defense minister. Turkey, on the other hand, complained that a Greek navy vessel had tried to ram one of its submarines and accused Greece of supporting the PKK (Kurdish Workers Party).

Greek support for the PKK caused a further downturn in relations with Turkey when, in 1999, it was revealed that the PKK leader Abdullah Öcalan, whom Turkey held responsible for the death of thousands of civilians and soldiers, had found shelter in the Greek embassy in Kenya. The fact that Öcalan had been in possession of a Greek Cypriot passport when he was captured added to the anger of the Turks (Gürbey 1999, 128). Turkish President Süleyman Demirel suggested

---

[9] The BSEC project was started in 1992 and encompasses Turkey, Bulgaria, Albania Greece, Romania, Moldova, Ukraine Russia, Georgia and Armenia (cf. Kramer 2000, 158). The primary aim is to promote cooperation in the fields of economy, technology, and environment, while cooperation might be extended into other areas in the future. The member-countries have agreed to establish a free trade zone by 2010.

[10] The creation of the force caused, however, some friction between Greece and Turkey, as both wanted the headquarters to be situated on their soil. In September 1998, agreement was reached according to which the headquarters is to be rotated among the member countries (cf. Uzgel 2001a, 63).

that Greece should be put on the list of countries supporting international terrorism. He added that Turkey would make use of its right of self-defense if Greece continued to provide support to the PKK. Greece would be given a last chance, he warned (Gürbey 1999, 123,132).[11] Simultaneously, the developments around the case of Öcalan had repercussions in Greece and led to the resignation of three ministers (see Clogg 2002, 230).

When earthquakes brought death and destruction to both countries in 1999, both people reached out to help their neighbor on the other side of the Aegean. An era of détente set in. In fact, efforts to defuse tensions and improve relations had already been underway. Turkish Foreign Minister Cem and his Greek counterpart Papandreou had met in June 1999 on the sidelines of a UN meeting (TDN, February 11th, 2002). Both countries had agreed on establishing committees to explore the possibilities for cooperation in areas such as economy, culture, tourism and multilateral cooperation in the region. In the aftermath of the earthquake in Turkey, Greece agreed to the release of EU funds to Turkey it had been blocking for years (Nachmani 2001, 84). A major stumbling block to closer relations was removed when the EU agreed to grant Turkey the status of a candidate country during its Helsinki summit in December 1999. Turkey committed itself to fulfilling the Copenhagen Criteria and undertaking efforts to solve existing bilateral disputes by peaceful means by 2004, and if bilateral dialogue failed, to refer them to the International Court of Justice (ICJ) thereafter. In return for Greek acceptance of Turkish candidacy, Turkey was reported to have accepted that a Cypriot membership could be realized prior to the resolution of the conflict.

In 2000, the Greek Foreign Minister Georgios Papandreou visited Ankara - the first time in 38 years that a Greek foreign minister had visited the capital of Turkey (Clogg 2002, 232; McDonald 2001, 142). Both countries signed a number of agreements on issues such as combating terrorism and safeguarding respective investments as well as on cooperation in the field of tourism and environmental protection (Reuter 2000, 57). When İsmail Cem later visited Athens, five further agreements were signed. They agreed on the creation of the aforementioned Southeast European Multinational Force within the framework of NATO's PfP as well as on the establishment of a Black Sea Bank of Cooperation and Development under the aegis of the BSEC, and signed a charter of good

---

[11] In October 1998, Turkey had threatened Syria with war if the latter did not stop providing support to the PKK and sheltering Öcalan (Gürbey 1999, 349). In order to reinforce the credibility of this threat, Turkey massed troops at the border and about 10,000 Turkish troops marched into Iraq. In the end, Syria agreed to extradite Öcalan and not to provide any further assistance to the PKK (Ibid., 356).

neighborliness, stability, security, and cooperation with an additional 21 SEECP (South East European Cooperation Process) states.

In January 2002, Turkish PM Ecevit revealed that foreign ministers of both countries had started discussing bilateral problems (Radikal, January 19th, 2002). Turkish officials affirmed that the foreign ministers had been discussing the problems and the modalities for solving them but added that negotiations had not started yet. Greek Foreign Minister Papandreou, too, acknowledged that they had been discussing bilateral problems to gain insight into the position of the other party and noted that they had not been negotiating sovereignty rights. Nevertheless, in 2002, both countries formally agreed to enter into dialogue on the possibility and conditions of referring the shelf-issue to the ICJ (Radikal, February 4th, 2002).

Nonetheless, despite this period of détente, their relations turned out to be anything but imperturbable and have encountered certain setbacks from time to time. In October 2000, the fragility of the rapprochement became visible when Turkey complained to the Alliance that Greek planes were using the airspace over demilitarized Aegean islands during NATO exercises and intercepted Greek bombers using these routes on their way to Turkey. Greece withdrew from the exercises in protest. The rapprochement process was further strained when Greece accused Turkey of increasingly violating Greek air space in the Aegean in May 2003 (cf. Kathimerini, May 17th, 2003). Apart from sending a note to the Turkish Foreign Ministry, Greece also informed the EU, NATO, and the ICAO about the incidents. However, the Turkish Foreign Minister, Abdullah Gül, tried to downplay the issue and asserted that serious work was being done between the two countries on Aegean issues and that the disputes in the Aegean should be resolved by 2004. Then, in June 2003, Greece claimed Turkish fighter jets had harassed a Greek civilian plane over the Aegean (NZZ, June 13th, 2003). While Greece spoke about Turkish provocations, Turkish military officials rejected the allegations. The Turkish press began to talk about Greek efforts to create tensions in order to block Turkey's entry into the EU. Still, the airspace issue did not impede an agreement on the new command structure of NATO. While Greece lost its Joint Sub-Regional Command, Turkey received the NATO Southern Region Air Command (cf. Hürriyet, June 13th, 2003). In contrast to the rhetoric used in former periods, Abdullah Gül stated that this should not be viewed as a victory for Turkey and that Greece would also make some contributions, adding that the command structure did not constitute a bilateral issue between the two neighbors. Avoiding inflammatory rhetoric, Gül maintained he would discuss the Greek allegations related to air space violations and attempts to internationalize the issue with his Greek counterpart. In the meantime, Deputy Prime Minister Mehmet Ali Şahin was of the opinion that the Greek note represented nothing new and that one should not make a camel out of a louse.

In May 2004, Turkish Prime Minister Erdoğan visited Greece. This visit was of symbolic importance as it was the first time that a Turkish prime minister visited the neighbor to the west in 16 years (cf. Hürriyet, May 7th, 2004). Erdoğan thanked Greece for its contributions to the decision taken by the EU to add the PKK and its successor organizations to the list of terrorist organizations and asserted that the Aegean should be turned into a "sea of peace". Greek Prime Minister Karamanlis, on the other hand, maintained that both countries had to win more in the future than what they had lost in the past.

While the general atmosphere of détente seems to be still in place, reports of air space violations and harassments of planes by the other side have recently started to appear more frequently in the newspapers on both sides. The Greeks have been claiming that Turkish fighter jets and patrol bots have repeatedly violated Greek air space and territorial waters in the Aegean (see for example Kathimerini, October 21[st,] October 27th, or November 11th, 2004). The Turks, on the other hand, have kept repeating that no such violations have taken place and that the problems can be attributed to disagreements related to the limits of national borders in the Aegean. On one occasion, Turkish Foreign Minister Gül even accused Greece planes of having harassed Turkish fighters while on NATO duty (TDN, November 11th, 2004).

Leaving aside these setbacks in this rapprochement process, the question arises of what factors have conditioned this change in the *quality* of Greek-Turkish relations and whether NATO has made a major contribution to this changing atmosphere. In retrospect, overall, the atmosphere of détente seems to have been conditioned by a number of factors unrelated to NATO. Analysts argue that the main factor behind the rapprochement process has been a change in Greek strategy. Such a view is maintained by Keridis, who draws attention to the changes in Greek policies towards Turkey and asserts that these changes reflect domestic developments in Greece (cf. Keridis 2001, 2). In the view of Keridis, forces in Greece favoring fiscal consolidation, the internalization of the economy, the shrinking of the state and a "redefinition of Greek identity within the framework of an open, multicultural European society" (Ibid., 8) are those forces that also want to extend the policy of deterrence vis-à-vis Turkey by a policy of engagement, which is expected to render Turkey more European (Ibid., 8, 14). In earlier periods, so Keridis, Greece "had no policy towards Turkey. Athens took a defensive stand, rejected all dialogue, and barricaded itself behind an over-legalistic self-righteousness. This was a safe strategy that placed Greek leaders at minimum risk and afforded them a convenient scapegoat against which to direct popular frustrations" (Keridis 2001, 12). Keridis holds that, along with the modernization of the political system and culture, Greek foreign policy has become more mature and flexible (Ibid., 11, 17). Greece has freed itself from the "siege mentality" and the "victimization syndrome" and has focused more on the opportunities that emerged in the Balkans with the end of the Cold War instead of being obsessed with risks and animosities. Keridis concludes that the Helsinki decision to grant candidate status to Turkey underscores that Greek foreign policy has become "more realistic, flexible, and imaginative- qualities that it has so rarely exhibited in the past" (2001, 17).

In a similar fashion, Triantoaphyllou asserts that Greek international relations entered a new era when Simitis won the elections of September 1996 (2001, 59). He speaks of a "paradigm shift" in Greek foreign policy (Ibid., 76). Larrabee, too, suggests that the improvement in Greek-Turkish relations has to a great extent been a result of changes in Greek policy (2001, 235-236). In his view, the Greeks realized that the policy of trying to isolate Turkey had been contra-productive. Instead, Greece changed its strategy to one of "Europeanizing" Turkey, hoping that the prospect of membership in the EU would induce Turkey to adopt policies more favorable to Greece. Greek desire to join the EMU, rendering it inevitable for Athens to reduce defense expenditures, is cited by Larrabee as a further reason for the change in Greek policy. In this regard, Auernheimer maintains that the issue of membership in the Monetary Union had been the most prominent political topic, pushing anything else into the background in the period preceding the elections of April 2000 (2000, 144). In a similar manner, Stivachtis holds that all major parties in Greece are interested in the welfare of the economy and in the exercise of fiscal discipline. A war with Turkey would obviously run contrary to these interests (Stivachtis 2001, 80).

In the view of Riemer and Stivachtis, another factor that gave rise to the rapprochement process was the circumstance that Greece and Turkey for the first time in 20 years were both ruled by politicians who displayed positive attitudes towards the US and NATO, encouraging the US to upgrade its efforts to contribute to an improvement in Greco-Turkish relations (Ibid., 560). The US and the Europeans were now willing to devote more attention to Greek-Turkish relations as they did not want any further disturbances between two allies when they had to deal with various serious crises in the Balkans. Neither did they want the Greek-Turkish antagonism to further destabilize Central Asia or the Middle East (Ibid., 561). Greece and Turkey, meanwhile, were familiar with the concerns of their bigger allies and did not want to cause further dissatisfaction in the West.

This was especially true for Greece which faced a growing power imbalance vis-à-vis Turkey and was dependent on the goodwill and support of its American and European partners (Ibid., 562). Turkey, in the view of Greece, had added to its strategic importance in the eye of the Americans, which might induce the Americans to support Turkish policy towards Greece. Thus, Greece wanted to recommend itself as a stabilizer rather than being a part of the problem.

Turkey, on the other hand, was facing serious domestic problems and did not want any further problems that might arise as a consequence of a renewed crisis with Greece. In addition, Turkey was aware that deterioration in relations with Greece more or less meant deterioration in relations with the European Union which Turkey wanted to join (Riemer & Stivachtis 2000, 563-564). Indeed, the *carrot* of EU membership has been one of the reasons why the Turkish government has pursued a resolution of the Cyprus issue so enthusiastically in recent months.

The earthquakes in 1999 are often cited as the point of return in Greek-Turkish relations. However, as already stated, first steps towards defusing tensions had already been taken before. Nonetheless, the earthquakes played an important role as the mutual exchange of aid led to a shift in the public opinion (cf. Bahçeli 2000, 457, 467). While public sentiments had usually been constraining the freedom of policy makers in shaping policies towards the opponent on the other side of the Aegean, the public this time favoured an improvement in bilateral relations. According to Bahçeli, the fact that the Davos process had been a "top down" initiative which was not embraced by the public was one of the factors rendering it unsuccessful (Ibid., 467). Today, the publics seem to be supporting the improvement in ties.

2.2) Areas of Dispute

2.2.1) Cyprus

The Ottomans captured Cyprus in 1571. In exchange for British support against Russia, they agreed to the occupation of the island by Great Britain in 1878. When the Ottomans entered World War I on the side of Germany and Austria-Hungary, the British annexed Cyprus, which then became a British crown colony in 1925 (cf. Varvaroussis 1979, 33).

Greece took the issue of Cyprus to the UN for the first time in 1954 and demanded self-determination (Varvaroussis 1979, 40). Greece obviously preferred UN forums to NATO, as it thought that a solution within the framework of the Atlantic Alliance would favor Turkey given the fact that Turkey was of greater strategic significance to the Alliance (Ibid., 40). It hoped that the non-aligned countries and the communists would support Cypriot demands for self-determination in the UN General Assembly. Turkey, on the other hand, initially wanted to see the status quo preserved. If, however, the principle of self-determination was to be accepted, the same rights had to be granted to the Turkish Cypriots (Ibid., 41).

When Great Britain turned out to be unwilling to accept self-determination or enosis - that is to say, union with Greece, the National Organization of Cypriot Fighters (EOKA-A) started an armed campaign against British rule in 1955 (Clogg 2002, 148).[12] In order to prevent enosis and to strengthen its position vis-à-vis Greece, Great Britain followed a strategy of involving Turkey in the conflict. In 1955, Great Britain invited Greece and Turkey to take part in a tripartite conference on the issue of Cyprus to be held in London. Before the scheduled end of the conference, an explosion in the courtyard of the Turkish consulate in Thessaloniki caused an uproar in Turkey (Stearns 1992, 28; see also Ierodiakonou 1971, 81). Anti-Greek riots in İzmir, İstanbul and Ankara followed. The Greek inhabitants of those cities were terrorized and their property looted.

The following years were marked by hostilities and fighting on the island. The parties involved failed to agree on a compromise for a further four years. Plans like that worked out by the British Governor of the island, Foot, or one drafted by British PM Macmillan failed to satisfy the demands of all parties involved (for the period see Ierodiakonou 1971). Finally, in February 1959, the Greek and Turkish foreign ministers succeeded in working out a compromise during their meeting in Zurich. A couple of days later, all parties, including the British and the two communities on the island, met in London to discuss the provisions of the agreement worked out by the two foreign ministers (cf. Ierodiakonou 1971, 206). In the end, all parties gave their consent to the agreement of Zurich and put their signatures under a final accord. The Greeks and Greek Cypriots had to disclaim enosis[13], the Turks had to give up the goal of taksim (partition) which they had

---

[12] After the EOKA was formed, the Turkish Cypriots established their own "resistance organization" called Volkan (cf. Fırat 2001b, 603). In 1958, the Türk Mukavamet Örgütü (TMT) was formed which secretly received financial and military assistance from Turkey. According to Fırat, the TMT had two goals, the first being a desire to establish itself as the main resistance force of the Turkish Cypriots. The second objective of the TMT was to achieve an escalation of inter-communal hostilities in order to show that the communities could not live together and that *taksim*- the partition of the island between Greek and Turkish Cypriots- was a necessary step (2001b, 606).

[13] However, according to Woodhouse (1982, 86-87), neither the Greeks nor the Greek Cypriots viewed this as the final stage in their campaign for enosis but rather hoped to fulfill the dream of

been pursuing since 1957 as an alternative to enosis, and the British had to cede the sovereignty over the island to the Cypriots except for two military bases.

Apart from outlining the new constitutional order on the island, the agreements of Zurich and London established Greece, Turkey and Great Britain as guarantor powers (Varvaroussis 1979, 35). In cases where the provisions of the treaties were violated, all guarantor powers were empowered to intervene in Cyprus' affairs, when possible jointly, if not, unilaterally, to restore the constitutional order envisaged in the treaties (Varvaroussis 1979, 35). Further, Greece and Turkey were allowed to station a certain number of troops, 950 and 650 respectively, on Cyprus. On August 16th, 1960, Cyprus became an independent state (see also Bağcı 1992, 137-138).

Archbishop Makarios, the president of the new republic, felt uneasy with the constitution of 1960 which left 30 percent of the seats in the parliament, 30 percent of government offices, and 40 percent of police posts to the Turks who only made up 18 percent of the population. Whereas he claimed the provisions of the Zurich and London accords were unworkable, the Turks asserted that, in fact, no real attempt had been made to implement these provisions (Tülümen 1998, 59). In 1963, he proposed 13 amendments to the constitution "which would have had the effect of drastically reducing the powers and representation guaranteed to the Turkish Cypriots" (Stearns 1992, 34). Turkish Cypriots rejected the amendments put forward by Makarios and withdrew from the government. In December 1963, violence broke out between the two communities. When the Turks, who were outnumbered by the Greek Cypriots, suffered severe casualties, Turkey signaled it would intervene on the island. A tripartite conference between Greece, Turkey and Britain did not bring about any results, and Britain looked to the US for help (Stearns 1992, 35). Undersecretary of State George Ball was sent to the region to mediate between the parties and to put forward proposals for a NATO peacekeeping force. However, his mission was to fail, too. In March 1964, a UN peacekeeping force was established on Cyprus. Yet, this force proved to be unable to put an end to the inter-communal fighting. When the Greek Cypriots took a decision envisioning the creation of an army, Turkey signaled it was intending to intervene on the island militarily (Sönmezoğlu 1995, 17; Harris 1972, 114). A harsh warning by US President Johnson, saying that Turkey could not count on NATO support in case of a Soviet intervention, helped to avert a pending Turkish

---

enosis in the long run. This view is shared by Uslu, who claims that the Greeks and Greek Cypriots viewed self-determination as a step towards the goal of enosis (Uslu 2000, 40).

military intervention on the island. Nevertheless, as the clashes continued, the Turkish air force carried out strikes against Greek Cypriot positions in August 1964.[14]

Greece and Turkey again came to the brink of war in 1967. When fighting between the National Guard joined by the police and the Turkish Cypriots left 27 Turks dead, Turkey carried out air strikes and threatened intervention again (Meinardus 1982a, 92-93; Woodhouse 1982, 187). The United States sent Cyrus Vance to mediate between the parties in an attempt to prevent a war. Manlio Brosio, the NATO Secretary-General, and a UN envoy joined the mediation efforts as well. In the end, a looming war could be prevented when Greece accepted Turkish demands to withdraw the Greek troops that had secretly arrived on Cyprus in 1964. Moreover, General Grivas, the leader of the EOKA, had to leave the island, and Greece accepted responsibility to compensate the victims of the fighting.

A split emerged between Athens and Nicosia during the Colonels' reign in Athens. Ioannidis, the chief of the junta, viewed Athens as the center of Hellenism (Clogg 2002, 163). Athens was the place where all decisions concerning the fate of Cyprus had to be taken. This was of course unacceptable to Makarios, who had by that time changed his stance and, for the time being, preferred independence to enosis. He thus asked Athens to withdraw all Greek officers from Cyprus. The Colonels, who were staunch anti-communists and had great misgivings about flirtations between Makarios and the communist bloc, responded by sponsoring a coup against Makarios carried out by the National Guard and the EOKA – B (Varvaroussis 1979, 47).[15]

However, the coup against Makarios and the installation of Nikos Sampson, who was known by the Turks for his atrocities against their compatriots, entailed an invasion by Turkish troops. Greece attempted a mobilization of its forces, which failed.[16] It shut down the allied headquarters in Saloniki, put all forces under

---

[14] In order to be able to launch air strikes, some units of the air force were withdrawn from the NATO integrated command. After the operation had been carried out, Turkey, in accordance with the appeals made by SACEUR Lemnitzer, assigned these units back to NATO (Harris 1972, 120).

[15] According to Clogg, Ioannidis undertook such an endeavor in order to consolidate his regime's hold on power (2002, 163). Given the student demonstrations and the naval mutiny of the previous year, Ioannidis might have sought to let the dream of enosis come true in order to raise the popularity of his regime (see also Woodhouse 1982, 205).

[16] As many officers had been thrown out of the armed forces given their opposition to the Colonels' regime, there were even not enough officers to command all the ships of the navy. Moreover, since the armed forces had been serving as a means of internal control, most of the tanks of the Greek army were stationed close to the capital instead of being deployed at the border

national command, and reinstated national control over all NATO installations (Varvaroussis 1979, 100). Two days after the invasion had begun, a cease-fire was brokered on Cyprus.

Representatives of Greece, Turkey and Britain met in Geneva on July 25th, 1974, to put an end to the war on Cyprus. The conference was adjourned five days later. The second phase of the talks started on August 8[th,] with the participation of the representatives of the two communities on the island. The negotiations ended in a deadlock on August 13[th,] and the second phase of the Turkish invasion started only two hours later (Meinardus 1982a, 199-201). The Turks captured about 36 percent of the territory before they ended their campaign.

The Turkish Cypriots declared the establishment of the "Turkish Federated State of Cyprus" in 1975. In 1977, an agreement was reached that a future solution should be based on the principles of bi-zonality and bi-communality. However, the communal talks did not bring about any solution and the Turks announced the establishment of the *Turkish Republic of Northern Cyprus* in 1983, recognized solely by Turkey.

In 1996, the incidents along the green line dividing the Greek and Turkish parts of Cyprus, leaving four Greek Cypriots and a Turkish Cypriot dead, caused renewed tensions on the Mediterranean island. Only a year later, tensions were to further increase when the Greek Cypriots announced plans for purchasing S-300 ground-to-air missiles from Russia (see Clogg 2002, 223, 229-230). Turkey threatened with air strikes if the missiles were to be deployed. These developments alerted Athens, as Greece and Cyprus had adopted a "unified defense doctrine" in 1993, according to which any aggression against Cyprus was to be regarded as aggression against Greece itself thus obliging it to come to Cyprus' aid in such a case.[17] Finally, Greece succeeded in persuading Cyprus that the missiles should be deployed on the island of Crete. According to Papacosma, one reason why Cyprus gave in to international pressure was the fear that the conflict might negatively affect the accession talks with the EU (Papacosma 2001, 216).

---

with Turkey (Meinardus 1982a, 178). In the end, a military regime had even failed to organize the mobilization of its troops. To put it in the words of Woodhouse (1982, 212), "[t]hus was exploded the delusion of the NATO powers that a military government could at least be relied upon for military efficiency."

[17] In fact, such a doctrine was adopted in 1963 (Meinardus 1982a, 171). The doctrine of the "common defense area Greece and Cyprus" envisaged Greek intervention in the event of aggression against Cyprus. In conformity with this doctrine, Greece had secretly sent a division of troops to Cyprus to deter a Turkish invasion.

The unified defense doctrine of Cyprus and Greece and the decision of the EU to start accession talks with the Greek Cypriots in 1998 induced Turkey and the Turkish Cypriots to extend their ties and promote integration in all fields. For this purpose, the Association Council was established in 1997 (Kramer 2000, 178). The EU decision not to grant candidate status to Turkey during its Luxemburg meeting of 1997 led to further frictions. While Turkey froze the dialogue with the EU and threatened to withdraw its application for membership, the Turkish Cypriots declined to continue the proximity talks under the auspices of the UN and abandoned the previously agreed formula of a federation, this time demanding the creation of a confederation of two states with equal political standing (Kramer 2000, 179, 194-199).

In 1999, inter-communal talks were resumed under the auspices of the UN (Axt 2003, 67). A second round of talks was held between January 31st, 2001, and February 8th, 2001 (McDonald 2001, 126). Towards the end of 2001 and the beginning of 2002, diplomacy between the two sides accelerated, thus giving rise to new hopes. In December, Clerides and Denktaş met for the first time after a four-year period (Hürriyet, December 5th, 2001). Both parties agreed to start direct talks on January 15th, 2002, to take place in Nicosia (NZZ, December 5th, 2001). On December 29th, 2001, Denktaş crossed the green line into Southern Cyprus for the first time since 1974 for a dinner with Clerides (Kathimerini, January 15th, 2002). In January 2002, the representatives of the communities started meeting up to three times a week to work out a solution. Unfortunately, these negotiations failed to bring about a solution.

The Secretary-General of the UN, Kofi Annan, put forward a new plan in November 2002 (cf. Axt 2003, 69-70). The plan envisaged the creation of two component states with far reaching autonomy, though united within the framework of a sovereign common-state.. However, the common-state was not to be a superior body. There was a clear division of competences between the common-state and the component states.[18] The presidency was to rotate between the representatives of the two communities. The component states would only sustain police forces while the common state would have a police force consisting of Greek and Turkish Cypriots. Only Greece and Turkey were allowed to station troops on the island. There should be a common citizenship as well as citizenship granted by the component states. Certain rights, however, were to be bound to the citizenship of the one or the other component state, thus limiting the freedom of

---

[18] Bahçeli states that Greek Cypriots reject the idea of a confederation as such a solution would provide for a separate Turkish state and yet endow them with veto rights and allow them to have a say in the affairs of the Greek Cypriots. Thus, a confederation as an outcome is regarded as worse than partition (2001, 221).

movement and settlement. The plan also envisaged territorial adjustments diminishing the share of the Turkish Cypriots from its current 36 percent to 28.5 percent.

Annan not only put forward a plan but also a timetable. First, a Foundation Agreement outlining the core provisions of the new constitution and drawing the new boundaries between the two constituent states was to be signed before the EU Copenhagen summit in December 2002 (cf. TDN, December 9th, 2002). The Final Agreement was to be signed before February 2003 and put to separate referendums on March 30th. Yet, the parties failed to adhere to the timetable once more, and the negotiations collapsed on March 11th, 2003. Only the Greek Cypriots signed the accession treaty to the EU in April 2003.

On April 23rd, 2003, the Turkish Republic of Northern Cyprus lifted travel restrictions between the south and the north (Kathimerini, May 23rd, 2003). Only a month later, *Kathimerini* reported more than 400,000 crossings by members of the two communities. Moreover, about 20,000 Turkish Cypriots were reported to have applied for Cypriot passports in order to become EU-citizens by May 2004. Meanwhile, Turkish Cypriots organized a number of demonstrations and called upon their leader to agree on a settlement with the Greek Cypriots in order to be able to join the EU by 2004.

Turkish Prime Minister Recep Tayyip Erdoğan, in an attempt to facilitate a resolution of the dispute on the island and to remove a stumbling block on the country's way towards EU membership, tried to persuade UN Secretary-General Kofi Annan to undertake a further attempt and call for a resumption of talks during a meeting on the sidelines of the World Economic Forum in Davos in January 2004. He assured US President Bush, too, that Turkey was interested in a quick resolution of the conflict on the island. Finally, the parties resumed talks under the aegis of the UN in February 2004. On February 14th, the parties were reported to have reached agreement on the further steps to be taken (NZZ, February 14th, 2004). Inter-communal talks would be resumed on February 19th on Cyprus. If the parties failed to agree on a solution, Turkey and Greece would enter the negotiations. In the case that a solution, nonetheless, did not emerge, the UN secretary-general would decide on matters the parties had failed to agree upon, and the final version of his plan would be put to referendums on both sides of Cyprus on April 24th.

Unfortunately, neither the talks in Cyprus, nor the negotiations in Buergenstock (Switzerland) with the participation of the Turkish and Greek officials who were later accompanied by the UN Secretary-General produced a solution. The parties could not agree on the fourth version of the UN plan. Taking into account the demands of the parties, the Secretary-General made amendments to his plan and put forward the fifth version. While the Turkish side was reported to have accepted the plan, Greek Cypriots rejected it. In accordance with the timetable agreed upon in New York, the plan was to be submitted to separate referendums on April 24th. Prior to the referendum, the Greek Cypriot President, Papadopoulos, argued that the plan satisfied Turkish demands while neglecting the wishes of the Greek Cypriot side and urged the Greek Cypriots to say "no" to the plan (Hürriyet, April 7th, 2004). He added that the plan did not constitute the last chance for the unification of the island and held that the Greek Cypriots would join the EU only a week after the referendum was scheduled to take place anyway. While Turkish Cypriot President Rauf Denktaş opposed the plan, and had thus refused to take part in the negotiations in Buergenstock, the Turkish Cypriot Prime Minister Talat wanted to accept the plan and join the EU on May 1st, 2004 with the Greek Cypriots.

In the end, as opinions polls had indicated, the majority (65 percent) of the Turkish Cypriots said "yes" to the plan while 75 percent of Greek Cypriots voted against it (Hürriyet, April 24th, 2004). The Greek Cypriot "no" was reported to have been conditioned largely by the provisions of the plan limiting the number of Greek Cypriots who would be allowed to return to their homes they had to leave after the 1974 invasion and allowing for the continued stationing of a number of Turkish troops on the island as well as by the fear that the financial burden of reunification might be too heavy (NZZ, April 24th, 2004).

After the plan had been rejected by the Greek Cypriot side, Turkish Cypriots and Turkey started to call upon the international community to put an end to the isolation of Northern Cyprus. Indeed, similar statements were made by EU and US officials claiming that the Turkish Cypriots would not be left out in the cold. In a first move, the Turkish Cypriots were allowed to attend Islamic Conference Organization gatherings as the "Turkish Cypriot State" (cf. Hürriyet, June 16th, 2004). While this neither meant recognition, nor changed the Turkish Cypriots' status as "observers", the Turks nevertheless welcomed the decision since the move was expected to allow for more contacts between the Turkish Cypriots and the representatives of other states.[19] A month later, the EU Commission proposed measures to allow for direct trade between the Union and Northern Cyprus and an aid package of about 259 million Euros to foster economic integration (NZZ, July 8th; 2004; Hürriyet 8th, 2004). The Greek Cypriot side argued that the measures

---

[19] Turkish Cypriot Foreign Minister Serdar Denktaş maintained that he had been able to meet the foreign ministers of 16 member countries on the first day of the conference while their requests for bilateral meetings had been turned down in a polite fashion previously (Hürriyet, June 16th, 2004).

proposed by the Commission could be interpreted as the first move towards recognition and warned to call the European Court of Justice should the Council of Ministers vote in favor of the measures. Nevertheless, a new EU regulation allowing for formal trade between the two parts of the island went into effect in August 2004. Simultaneously, the Turkish Cypriots decided to lift the ban on exports from the south, while trade relations are reportedly intended to be gradually initiated (TDN, August 27th, 2004). Most recently, the Turkish side hailed a decision by the Parliamentary Assembly of the European Council to allow Turkish Cypriots to be represented separately from the Greek Cypriots as a right move towards ending the isolation of the Turkish Cypriots (cf. TDN, October 6th, 2004). Previously, they were represented under the Greek Cypriot umbrella and were only allowed to watch general assembly meetings from the audience lounge. Meanwhile, the US administration is said to be intending to allow international flights to the Turkish Cypriot port of Ercan (TDN, November 8th, 2004). Given the US decision to recognize the FYROM simply as "Macedonia," which caused uneasiness in Greece, the Greek Cypriots are reported to be worrisome that the recognition of the Turkish Cypriot state might follow (TDN, November 8th, 2004; Hürriyet, November 8th, 2004).

On the other hand, as the date when the EU Council is to decide on the question as to whether accession talks with Turkey should start or not is getting closer and closer, the Greek Cypriots have been reiterating their claim that Turkey must recognize the country if it wanted to join the Union. In a similar fashion, during a press conference held in the aftermath of the EU Troika meeting, the Dutch foreign minister, Bot, stated that it would be better for Turkey to recognize the Greek Cypriot state before December 17th (Hürriyet, November 25th, 2004). However, the Turkish government has repeatedly declared that it had no obligation to recognize the Greek Cypriot administration (Kathimerini, November 11th, 2004). While Greek Cypriots are unlikely to veto the beginning of accession talks, they are likely to use the issue as a bargaining chip to pressure more concessions from the Turkish side during the negotiation process.[20]

---

[20] In May 2004, Karamanlis had declared that Greek Cypriots would not cause any difficulties with regard to Turkey's EU bid (Hürriyet, May 7th, 2004). After a meeting with the Greek Cypriot President, Papadopoulos, in mid-November, Karamanlis asserted that a country that wanted to join the EU had to understand that there were certain rules of behavior (cf. Kathimerini, November 11th, 2004). Papadopoulos meanwhile claimed that Turkey would have no choice but to recognize Greek Cyprus once the accession talks had started (see also Hürriyet, November 15th, 2004). Nevertheless, both leaders' statements fell short of saying they were considering vetoing the beginning of talks with Turkey. Back in September, the Greek Foreign Minister had declared that Greece would support Turkey's EU bid even if all other 24 members opposed it (Hürriyet, September 22nd, 2004). Moreover, in October 2004, during a visit to Nicosia the British Minister

No matter what form the decision taken by the heads of state of the 25 EU member-states takes on December 17th, 2004, inter-communal relations seem to have achieved a different quality more conducive to permanent peace and tranquility. The fact that the UN Secretary-General called for a 30 percent reduction in the number of UN peace-keeping forces on the island (TDN, October 25th, 2004) and two British contractors were tasked with clearing 8 minefields in the Greek part of Nicosia (Kathimerini, November 16th, 2004) is exemplary to this atmosphere of détente.

2.2.2) The Delineation of the Continental Shelf

The major contentious issue related to the Aegean is the delineation of the continental shelf. Ever since a crisis erupted in 1973, both countries have failed to produce a formula for delimitation. While Greece claims that, in line with the provisions of the 1958 Geneva Convention on the continental shelf and the 1982 UN Convention on the Law of the Sea, the islands in the Aegean are entitled to have a shelf, Turkey asserts that the islands are located on the extension of the shelf of the Anatolian mainland and lays claim to the area extending to the middle of the Aegean Sea (Riemer 2000, 125-126; Axt & Kramer 1990, 14-18). In the view of the Turks, the conventions cited by Greece have no jurisdiction in this case because Turkey is not a party to the treaties. Greece, on the other hand, asserts that the fact Turkey is not a party does not make a difference since the treaties embody codified customary law. In contrast to the Turkish claim that the line dividing the shelf must be the median line dividing the Aegean, Greece claims that the division line must run between the Eastern Aegean islands and the Turkish mainland.

As mentioned above, the first serious dispute related to the delineation of the Aegean continental shelf erupted in November 1973. The oil crisis of the early 1970s had forced both countries to intensify their search for other sources to meet their demands for oil (see Clogg 2002, 173; Meinardus 1982a, 289). Hence, both had turned their attention to the continental shelf of the Aegean Sea where oil reserves were thought to exist. After carrying out drillings in 1973, Greece - which had issued licenses for explorations already in 1961 (Meinardus 1982a, 291) - had announced that oil and gas reserves had been found in the vicinity of the Greek island of Thasos (Coufoudakis 1985, 199). Turkey was to follow and issued licenses for carrying out explorations as well as for the exploitation of found reserves. However, the explorations were to be carried out in areas Greece

---

for Europe, Denis MacShane, maintained that he had been assured by the Greek Cypriot side that they would not veto the opening of accession talks with Turkey (TDN, October 22nd, 2004).

had laid a claim upon. A war threatened to break out when, in May 1974, Turkey sent the research vessel Çandarlı, escorted by military vessels and aircraft, into the disputed area to begin with soundings. Greece put its armed forces on alert and declared it would view it as a *casus belli* if Çandarlı carried out soundings in waters claimed by Greece. The 1974 Aegean crisis was to be overshadowed and pushed into the background by the developments on Cyprus (see also Meinardus 1982a, 312).

In 1975, Greece offered Turkey to refer the shelf issue to the ICJ in The Hague (Ibid., 230). Turkey accepted the Greek proposal, and experts from both countries started to work on the modalities of referring the issue to the Court, as the procedures of the ICJ require both countries' agreement to jurisdiction by the ICJ. In the end, the experts failed to work out a compromise, and this barred the way to a settlement via international jurisdiction.

The events of 1974 were to repeat themselves in 1976 when Turkey sent another vessel (Sismik I) into Aegean waters to explore the continental shelf. As Greek protests did not bring about any results, Karamanlis called the Security Council of the UN for help and asked the International Court of Justice to take up the issue and prescribe interim measures for the protection of Greek interests. The leader of the Panhellenic Socialist Movement (PASOK), Andreas Papandreou, meanwhile, demanded that the Greek navy should sink the Turkish ship. The UN Security Council warned both countries not to use force and advised them to solve the issue through bilateral negotiations and to consider the possibility of referring the issue to the ICJ (Axt & Kramer 1990, 18-20). The ICJ, on the other hand, said it had no jurisdiction in the case on the grounds the issue had been taken to the court by Greece unilaterally. The Court also refused to take any interim measures (Varvaroussis 1979, 231-232; see also Clogg 2002, 128; Stearns 1992, 136). Nonetheless, in the same year, both countries agreed to enter into bilateral negotiations on the shelf issue and accepted - in principle - that the issue could be referred to the ICJ for a resolution. In the "Bern Declaration" both countries stated their willingness to abstain from any activities which might prejudice a solution to the shelf issue (Meinardus 1982a, 292, 383) and concurred to keep the talks confidential (see also Kürkçüoğlu 2000, 47). These talks were suspended when the Papandreou government came into office in Greece in 1981.

The latest crisis of 1987 erupted when Greece granted new licenses for carrying out drillings in areas outside its territorial waters (Riemer 2000, 117).[21] Turkey

---

[21] According to Fırat, on February 27th, 1987, the Greek deputy foreign minister informed the Turkish ambassador that Greece no longer recognized the Bern Declaration and might thus carry out drillings in the Aegean (2001d, 112-113). Fırat argues that Greece might have been

responded by sending a research vessel escorted by military ships to the area (see also Clogg 2002, 188).[22] Greece, once more, put its forces on high alert and sent the Greek navy into the area. On March 27th, 1987, the Greeks declared they would prevent the Turkish vessel from carrying out exploration activities (Axt &Kramer 1990, 40). The crisis ended after mediation by the United States and other NATO member-countries. Turkey accepted that its vessel should not drill outside the Turkish territorial waters, and Greece agreed to abstain from drilling in disputed areas.

2.2.3) The Breadth of Territorial Waters

For the time being, both Greece and Turkey claim a six-mile territorial sea in the Aegean. Greece, however, maintains that the 1982 UN Convention of the Law of the Sea (UNCLOS) allows for the extension of its territorial waters to 12 miles. Simultaneously, Greece states it has no intention to do so for the time being. Greece had already in 1974 announced it would extend its territorial waters to 12 miles, but refrained from doing so (see Kürkçüoğlu 2000, 51-55).

If Greece indeed implemented the 12-mile regulation, Greek territorial waters would rise from currently 35 percent to 63.9 per cent, the breadth of Turkish territorial would only slightly change from 8.8 to 10 percent, while international waters would decline from 56.2 to 26.1 percent of the Aegean Sea (cf. Stearns 1992, 139). Moreover, an extension of Greek territorial waters would also prejudice the outcome of the shelf dispute, since the shelf claimed by Greece lies within the 12-mile zone (Bahçeli 2000, 459-460). Given the international practice according to which sea determines air, Greece would also be able to claim a 12-mile airspace. Claiming that such a move would turn the Aegean into a Greek lake and cut Turkey off from the international waters restricting the freedom of movement of its military, Turkey declared that a unilateral extension of Greek

---

deliberately trying to raise tensions just before Turkey was to file its admission application to the EC. Meanwhile, Greece had in fact been carrying out such drillings since the beginning of the 1980s. However, Turkish Prime Minister Özal, who was interested in improving ties, had refrained from informing the public about such incidents. The general staff and the foreign ministry were, however, not content with the concessionary policies of Özal vis-à-vis Greece. When the new crisis erupted, Özal was abroad and the Turkish response was fashioned by the general staff and the foreign ministry, and this time the Greek move would not be ignored and a Turkish ship was sent to the region.

[22] Güven Erkaya states that Turkey had received intelligence that Greece had granted an exploration license to a Norwegian ship (Baytok 2001, 128). Thereupon, the Turkish research vessel Sismik anchored in the Aegean and a plane and a ship were sent into the area where the Norwegian ship was reported to be However, the prior intelligence could not be verified. Thereafter, tensions defused.

territorial waters to 12 miles would constitute a *casus belli* (see also Riemer 2000, 123-124, Axt & Kramer 1990, 24; Stivachtis 2001, 42-44). Thus, when the Greek parliament ratified the 1982 convention in 1994, the Turkish parliament empowered the government to take all necessary steps, including military action, to protect the rights of Turkey in the Aegean (Bahçeli 2000, 460).

2.2.4) Problems Related to the Airspace

Whereas in international practice the boundaries of national airspace are identical with those of territorial waters, Greece claims 10 miles of airspace for islands in the Aegean while the breadth of its territorial waters amounts to 6 miles (cf. Stearns 1992, 140). Greece asserts this is necessary for the protection of the islands and states that the 10-mile zone was established by a presidential decree back in 1931 and that Turkey did not object to this regulation before 1974. Turkey, on the other hand, refuses to accept a 10-mile airspace for the islands, and Turkish military jets continuously challenge the 10-mile boundaries resulting in mock dogfights between the aircrafts of both NATO members (Ibid., 141).[23]

The control of civilian air traffic also became a contentious issue after the Turkish invasion of Cyprus. The areas for the control of the air traffic were decided upon by the International Civilian Aviation Organization (ICAO) in 1952. According to the decision taken by the ICAO, the boundaries of the Flight Information Region (FIR) Athens extend over the entire Aegean and intrude into the Turkish airspace (Meinardus 1982a, 343). These division lines approved by the ICAO were adopted by NATO in 1964 for its military command structure. Neither the ICAO nor the NATO regulations constituted a problem until the emergence of the conflict related to the delineation of the continental shelf in 1973 (Ibid., 343). In the same year, Turkey referred the issue to the ICAO and demanded a revision of the boundaries. However, it was after the invasion of Cyprus that Turkey issued a Notice to Airmen (NOTAM 714) declaring all planes had to inform the flight-control in İstanbul on crossing the median line of the Aegean. Turkey claimed this was a necessary measure adopted against the danger of attacks by the Greek Air Force. In response, Greece issued NOTAM 1157 and declared the whole Aegean to be a dangerous zone. As a consequence, all air corridors over the Aegean were closed and the air traffic came to a standstill. Mediation by the ICAO and bilateral negotiations did not bring about a solution. However, in the face of costs arising from the dispute, Turkey withdrew its NOTAM in 1980 and tacitly recognized the

---

[23] NATO recognized a Greek air space of 6 miles in 1960 (MC 66/1/60), and Greece acknowledged this tacitly during the negotiations on Greek reintegration (Meinardus 1982a, 353; see also Güldemir 1986, 83).

existing FIR-lines. Greece quickly followed and withdrew NOTAM 1157, whereupon the Aegean airspace was reopened to civilian air traffic (Meinardus 1982a, 346-347).

In the Turkish view, FIR responsibilities do not cover military air traffic, and thus Turkish military planes are not obliged to inform the air-control in Athens about their position and flight plans on crossing the FIR line (Meinardus 1982a, 349-351; Stivachtis 2001, 46). Moreover, Greek authorities are accused of abusing FIR provisions to constrain the freedom of movement of the Turkish Air Force in the region (Baytok 2000, 156). In fact, until 1980, Turkish military planes had been submitting their plans upon entering FIR Athens (cf. Bölükbaşı 1992, 40). Yet, the Turks abandoned this practice on the grounds that Greek authorities had been restricting the freedom of movement of the Turkish Air Force consistently demanding changes in the timing or in the flight plans, setting up additional national air corridors and control zones in the international airspace, and sometimes claiming that the exercises could not take place for reasons of safety concerning civilian air traffic (Ibid., 40). Greece, on the other hand, asserts that ICAO provisions do encompass military flights and demands that Turkish jets identify themselves and inform Athens about their flight plans. In this context, Greek jets have regularly been intercepting Turkish jets in Aegean air space and asking for identification since 1980.

A further point of discontent arose when, in 1989, Turkey declared that it was responsible for "search and rescue" operations in the area reaching to the middle of the Aegean Sea (Papacosma 2001, 207-208; Axt & Kramer 1990, 68). The area of responsibility claimed by Turkey overlapped with the part of the continental shelf Turkey laid a claim to, and contradicted the regulations of the ICAO, according to which the areas for "search and rescue operations" correspond to the boundaries of the flight information regions.

2.2.5) Command and Control Rights

Before Greece left the military arm of NATO, a Greek commander and a Turkish commander subordinated to an American officer were in charge of the affairs of the NATO headquarters in İzmir. The Greek commander was responsible for the airspace over the Aegean - except for the Turkish national airspace. When the Greek officers left İzmir in 1974, the command was transferred to a Turkish officer who was subordinated to the Commander-in-Chief, Allied Forces Southern Europe (CINCSOUTH), in Naples, Italy (Stearns 1992, 70; Meinardus 1982a, 408). The sharing of command and control rights was to emerge as a contentious issue during the negotiations on the reintegration of the Greek forces.

Turkey now demanded that the boundaries of the command and control areas over the Aegean be redrawn in a manner extending the Turkish zone of responsibility to the middle of the Aegean Sea.[24] Greece, on the other hand, wanted the *status quo ante* 1974 to be restored. The issue could not be resolved despite mediation efforts undertaken by the Supreme Allied Commander in Europe (SACEUR), General Haig. Haig proposed three plans for resolving the issue but none gained the acceptance of both parties.[25] When General Haig left office, his successor, General Rogers, continued the search for a solution. The second plan put forward by Rogers was finally accepted by the parties involved.

The Rogers plan, in fact, did not actually provide for a solution as the thorny questions were left out. The decision on the boundaries of the command and control areas was to be taken after Greece's reintegration had taken place and a new NATO headquarters had been established in Larissa. The negotiations started in 1980 but ended in a deadlock in 1981. Greece, at that time, refused to set up a new command. In the Greek view, an agreement on the boundaries of the command and control areas had to be reached prior to the establishment of the new command (cf. Stearns 1992, 69-71; Papacosma 2001, 206). The issue could not be solved for almost two decades, and the new command was only established after the NATO member-states concurred on a new command structure for the Alliance in 1997.

This dispute between Greece and Turkey was solved by the introduction of "no-boundary commands" (International Institute for Strategic Studies 1998, 43-44). This new arrangement was to allow only for a functional coordination between the commands, while there should no longer be any geographically delineated areas of responsibility (Interviews in Brussels, July 2001). The new agreement envisaged that the two countries share control over NATO flights in the region (Papacosma 2001, 214) and the relationship between the command in Larissa and the other one in İzmir be determined by the headquarters in Naples when a crisis arose- "the worst time to do so" (International Institute for Strategic Studies 1998, 44). The new arrangements emphasizing multi-nationality allowed Greek and Turkish officers to work in the respective headquarters on the other side of the Aegean (Kathimerini, January 22nd, 2003).[26]

---

[24] According to Erkaya, in 1973, Turkey had already achieved the inclusion of the problem related to the sharing of command and control rights in the Aegean into a NATO document which listed the problems requiring a solution within the Alliance (Baytok 2001, 160-161).

[25] For the provisions of the Haig plans see Meinardus 1982a, 410-411.

[26] According to Roper, the command arrangements in Southeast Europe had been inefficient, and the two sub-regional commands in İzmir and Larissa had been maintained for "purely political reasons" (Roper 1999a, 93). A NATO official acknowledged that different views on the effectiveness of the new structure existed and claimed those who assert the structure was

In June 2003, the Alliance decided on a further restructuring entailing an additional cut in the number of existing headquarters. The new command structure adopted envisaged the closing down of the joint sub-regional command in Larissa (cf. Kathimerini, June 13th, 2003). The Combined Air Operations Center (CAOC) was to remain in Larissa and to answer to the air command center in İzmir. Moreover, a NATO Maritime Interdiction Operational Training Center would be set up in Greece, tasked with anti-terrorism operations in the Mediterranean. While the Greek government claimed the new structure was satisfactory, the spokesman of the party in opposition, New Democracy, blamed the defense minister for having given up Greek rights and responsibilities to the benefit of Turkey. Greece had originally insisted, "what holds for Greece should also hold for Turkey" (Kathimerini, May 19th, 2003) and wanted the air command AIRSOUTH to be on Greek territory. If the command was not to be located on Greek soil, it should remain in Naples. In the end, Turkey was given AIRSOUTH.

2.2.6) Militarization of Greek Islands

As early as 1960, Greece started to militarize a number of islands in the Eastern Aegean. Turkey declared that the militarization of these islands was illegal and posed a threat to the security of its western coast. Greece, on the other hand, referred to other treaties as the legal basis of the militarization and described it as a necessary measure to counter the threat posed by Turkey (Meinardus 1982a, 360-364). For example, with regard to the islands of Lemnos and Samothrake, Turkey cites the Lausanne Treaty as the legal basis according to which these islands must not be militarized. Greece, on the other hand, asserts that the restrictions on the militarization of these islands were lifted with the Montreux Convention of 1936, which stated that the straits could be remilitarized again (Axt & Kramer 1990, 29-30).[27] Concerning the islands of Lesvos, Chios, Samos and Ikaria, Greece offers the view that the Lausanne Treaty called only for partial demilitarization while Turkey demands total demilitarization. The Dodecanese Islands were ceded to Greece in 1947, but they had to remain demilitarized except for security forces to maintain internal order (Ibid., 30). Greece maintains that Turkey cannot protest against the militarization of these islands because it is not

---

ineffective failed to recognize that NATO's sole function was not to fight a war. NATO's approach to stability and peace was comprehensive, and to have a headquarters on their soil constituted a value in itself for member-countries (Interviews in Brussels, July 2001).

[27] Given the atmosphere of détente between the two countries when the Montreaux Convention was signed, the Turkish Foreign Minister of the time, Aras, stated that the restrictions regarding the militarization of Lemnos and Samothrake had been lifted with the Convention. Turkey claims that the statement by Aras constituted a goodwill gesture towards Greece and does not change the fact that Convention provisions cannot be applied to the islands in question (cf. Bölükbaşı 1992, 42).

party to the Paris Treaty (Ibid., 31). In the meantime, Greece points to Article 51 of the UN Charter and claims that the militarization of the islands has to be viewed as a defensive measure in line with the right to self-defense (Stivachtis 2001, 65).

Particularly the status of the island of Lemnos has been causing discontent between the two countries. As early as 1958, Turkey refused to accept NATO funding for an extension of the then civilian airport in Lemnos (Meinardus 1982a, 367). In 1982, when Turkey objected to the island's inclusion in NATO's military maneuvers and NATO Secretary-General Luns told NATO bodies not to include the island in exercises, Greece, in protest, annulled its participation in the scheduled NATO exercise Apex Express (see also Axt & Kramer 1990, 30). Ever since, Turkey has opposed the inclusion of the island in NATO maneuvers and in NATO force planning. Greece, on the other hand, refused to participate in the maneuvers in the Aegean excluding Lemnos.

One should add that after the hostilities of 1974 on Cyprus and, in Turkish argumentation, as a response to the militarization of the Greek islands, Turkey established the Aegean Army in 1975 (Meinardus 1982a, 371). The Aegean Army is not assigned to NATO and is fully under national command. Whereas Turkey refers to it as a training army, Greece perceives it as an invasion force posing a serious threat to the security of the Greek islands in the Aegean.[28]

2.2.7) The Status of Certain Islets in the Aegean

In January 1996 a new crisis erupted between Greece and Turkey over the status of an islet called Imia by the Greeks and Kardak by the Turks. Greece claimed the islet was ceded to Greece in 1947 with the Dodecanese Islands which had belonged to Turkey before they were ceded to Italy in 1932 (NZZ, January 31st, 1996). Turkey claimed that the 1932 treaty was not valid as it was not registered with the League of Nations and the status of the "Kardak rocks" was unclear and thus offered to enter into negotiations. Greece, however, claimed that the islet was a part of Greek territory and there was nothing further to negotiate.

After a Greek mayor had hoisted the Greek flag, two Turkish journalists working for the daily *Hürriyet* were flown to the islet to replace the Greek flag by a

---

[28] In the view of Mackenzie, the fourth army "is very much an army on paper, used for training purposes and devoid of striking power" (1983, 13). In his view, the army serves as a means of psychological warfare and Turkey could capture the Greek islands across the Turkish coast in the event of a crisis even if the fourth army did not exist. Note that, according to Nachmani, the Turkish forces along the Aegean coast are kept with 80 – 100 percent readiness while elsewhere the usual level is 50 percent (Nachmani 2001, 73).

Turkish one (NZZ, January 31st, 1996). Greece responded by sending a navy vessel to the area in order to have the Turkish flag removed (NZZ, January 30th, 1996). The tensions escalated when the Turkish navy sailed out into the Aegean, and both countries put their armed forces on high alert. Greek and Turkish forces landed on nearby islets to reinforce their claims to the tiny rocks.[29]

US mediation helped to prevent a war once more (cf. NZZ, February 1st, 1996, February 2nd, 1996). According to the US mediator, Holbrooke,[30] who had played an important role in defusing tensions, the Turkish forces had intended to capture the contended islet where Greek forces had positioned themselves. Such a move would have probably caused Greek retaliation. When President Clinton was informed about the developments, he intervened by telephone and tried to mediate between the parties. Finally, the disputants could agree on withdrawing their forces and returning to the status quo ante. US planes were to monitor the withdrawal of the Greek and Turkish navies and inform both parties about the actions of the other side.

Whereas Greece claims that Turkey challenges its territorial integrity by questioning the status of certain islets and rocks in the Aegean, in the Turkish view, the status of these islets is unclear, as they have not been explicitly ceded to Greece by international treaties (cf. Stivachtis 2001, 47). Thus Turkey speaks of Greek expansionism. According to Fırat, after the 1982 Convention had come into effect in 1994, Greece adopted two strategies to assert sovereignty over the islets and to prove that these islets were entitled to have their own continental shelf (Fırat 2001e, 469-470). On the one hand, Greece tried to secure the inclusion of these islets in NATO exercises, which would indirectly affirm the Greek possession of the islets. In the meantime, Greece tried to settle Greeks or EU citizens on the islets, because Article 121, § 3, of the 1982 Convention stated that islets which were not habituated or had no independent economic life were not entitled to have an economic zone or their own continental shelf. Turkey, on the other hand, wanted to prevent such an outcome and put forward its thesis of "gray areas" claiming the status of a number of small islands, islets, and rocks were not

---

[29] Meanwhile, Turkey was reported to have reinforced its troops on Cyprus by sending about 40 additional tanks to the island (NZZ, February 2nd, 1996).

[30] Holbrooke warned, too, that constant tensions in the region could even lead to a dissolution of NATO (NZZ, February 2nd, 1996). He added that both countries preferred to deal with each other through Washington, and maintained (TDN, February 2nd, 1996): "In the end, neither side would give the other a guarantee they wouldn't do something again in the future.[...] But each side was willing to say to the United States they wouldn't do something in the future...and we conveyed it. The United States is holding escrow these two agreements that they will go back to the status quo ante [...]."

clear and implying that they might belong to Turkey as well. In 1997, Turkish President Demirel claimed that the status of 130 islands was debatable (Clogg 2002, 223).

2.2.8) Minorities

The Lausanne Treaty of 1923 included provisions for the exchange of populations living in Greece and Turkey (Meinardus 1982a, 489-493; Clogg 2002, 99-100). Almost 380,000 Turks and 1.1 million Greeks had to leave their homes and resettle in their *home countries*.[31] Only the Greek minority in İstanbul and the Turks living in Western Thrace were excluded from the exchange. In addition, the islands of Bozcaada and Gökceada populated mainly by Greeks were given to Turkey.

As the relations between the two countries continued to deteriorate, their respective minorities were to suffer under measures taken to retaliate against the actions of the opponent. Both parties linked the welfare of the respective minorities to the developments on Cyprus and to the treatment of their minority by the other side. Baskın Oran's claim, according to which the Turkish Foreign Ministry learned of discriminatory measures adopted by the General Directorate of Foundations in Turkey (Türkiye Vakıflar Genel Müdürlüğü) against Greek foundations and decrees by the education ministry violating the rights of the Greek minority only after Greece had already adopted retaliatory measures against the Turkish minority in Western Thrace is illustrative of the principle of reciprocity that characterizes the treatment of the minorities (Oran 2001, 308).

As mentioned previously, the failure of the governments to solve the Cyprus issued led to anti-Greek riots in Turkey in 1955. Later in 1964, Turkey responded to a new crisis on Cyprus by rescinding a degree from the year 1930 which granted certain special rights to the minority in İstanbul (Meinardus 1982a, 520; Oran 2001, 308). Moreover, 12,000 Greek citizens were extradited on the grounds that they constituted a security threat. When a new crisis erupted in 1967, Turkey extradited a number of Greek teachers and closed down 9 schools run by the Greek minority. In addition, Turkish deputy headmasters were assigned to all remaining schools to supervise their activities (Meinardus 1982a, 551). Turkish regulations stating that the Ecumenical Patriarch must be a Turkish citizen chosen by the Holy Synod from a list of candidates approved by the Turkish government constitute another source of discontent (Stivachtis 2001, 50). Greece also accuses Turkey of trying to eradicate the Greek minority living on the islands of Gökceada

---

[31] The numbers provided by both authors differ slightly. Meinardus refers to 434,000 Turkish and 1.35 million Greek emigrants (1982a, 492).

and Bozcaada (Stivachtis 2001, 71). For example, the installation of an open prison on the island of Gökceada was viewed by Greece as a measure devised to force the Greek population to leave in violation of the provisions of the Lausanne Treaty (Ibid., 71). Overall, the number of Greeks living in Turkey decreased from 100,000 in the 1950s to 1,000 in 2000 (Clogg 2002, 205).

On the other hand, Greece adopted measures detrimental to the fate of the Turkish minority in Western Thrace. Greece was accused of adopting discriminatory measures against the minority in Western Thrace, restricting property rights, and annulling their right to elect religious leaders. This view was reinforced by a report of the US State Department published in 1991 that maintained that the Muslim minority was subjected to economic and social discrimination (Clogg 2002, 206). Members of the Turkish minority in Thrace who settled in Turkey or went to Turkey to study there were stripped of their Greek citizenship (cf. Tülümen 1998, 47).[32] The freedom of movement of the minority was restricted. Furthermore, Tülümen – a retired Turkish diplomat - accuses Greece of adopting restrictive measures in the field of education and attempting a forced assimilation of the members of the Turkish minority (Ibid., 47-48). In cases where assimilation fails, attempts are taken to force the people to emigrate (Stivachtis 2001, 54). The fact that the Greek police turned away and looked in the other direction when a Greek mob attacked a number of Turks and destroyed 200 shops in 1990 was to cause further disgruntlement (Vamik & Itzkowitz 1994, 161). In sum, Turkey alleges Greek human rights violations in Western Thrace. Greece, on the other hand, asserts that Turkey supports the minority in Western Thrace to weaken the country from within and fears the neighbor might one day even try to annex the region citing human rights abuses as a pretext (Stivachtis 2001, 68).

However, in recent years, the Greek government has adopted measures to improve the situation of the minority in Western Thrace (cf. Bahçeli 2000, 464). Economic measures intended to raise the living standards in Western Thrace were implemented. In addition, the Simitis government abolished the provisions of the constitution depriving members of the Turkish minority of their Greek nationality when they left the country. Above all, Greece, which had previously declined to acknowledge that a Turkish minority resided in Western Thrace, changed its policy and recognized the minority in Thrace as a Turkish one (see also Nicolaidis 2001, 252). In 1996, three members of the Turkish minority were elected to the Greek parliament (Papahadjopoulos 1998, 28). Meanwhile, Turkey refrained from

---

[32] According to the Turkish Foreign Minister, Abdullah Gül, 60,000 members of the minorities in Greece were denaturalized between 1995-1998 as a result of Greek regulations, Turks constituting the majority thereof (Hürriyet, June 8th, 2004). The Foreign Minister added that the regulation in question was abandoned in 1998.

unfriendly acts against the Greek minority. In November 2003, Turkey signaled that the Greek Orthodox theological school in Heybeliada, which was closed down in 1971, might be reopened soon and urged the Greek side to adopt similar measures and improve the situation of the Turkish minority in Greece (TDN, November 12th, 2003).[33]

---

[33] As Turkey is preparing to revise the "National Security Policy Document", the foreign ministry is reported to have proposed that the statements related to the theological school on Heybeliada describing its reopening as a threat to the country's security be deleted from the document (Hürriyet, November 28th, 2004).

3) THEORIZING ON INTERNATIONAL INSTITUTIONS: (NEO)REALISM vs. NEOLIBERAL INSTITUTIONALISM

As stated in the introduction to this work, neorealists and neoliberal institutionalists continue to disagree on the prospects for international cooperation and on the question as to whether international institutions - Keohane defines institutions as "persistent and connected sets of rules (formal or informal) that prescribe behavioral roles, constrain activity, and shape expectations" (1989, 3) - can contribute to peace and stability in the international arena. This section aims to provide an insight into this debate and answer the question regarding the kind of functions international institutions might perform. A further issue that will be dealt with is the role played by international institutions in domestic democratization processes. This debate of institutional *merits* in general and their impact on democratization in particular is expected to enable us to narrow down the scope of analysis and focus on a number of *functions* against which NATO's *performance* could be assessed in the following sections.

International Institutions and International Cooperation

Realists view international politics as a struggle for security and survival (cf. Mearsheimer 1994/95, 9-14). International politics lacks the kind of hierarchy that characterizes domestic politics. There is no world government tasked with guaranteeing the security of states and undertaking punitive actions against aggressors. It is an anarchic self-help system (see also Waltz 1979, 88-102). Hence, every state has to take care of itself and guarantee its own survival. Furthermore, states always face a high degree of uncertainty. They can never know whether others' designs are benign or not. Even if they knew the motives of other states with a 100 percent certainty, nobody can guarantee that these motives will not change over time. Given this insecurity arising from the anarchic state of the system and the uncertainty felt about the motives of other states, distrust prevails in states' relations with each other. Today's friends can be tomorrow's enemies, and *vice versa*.

All these features of the international system set certain constraints on the possibility of interstate cooperation and on the scope of such cooperation if it occurs (Waltz 1979, 105-106). States as rational actors have to secure their relative power positions in the system and limit the degree of their dependency on others. It is in their outmost interest to prevent others from achieving disproportional gains in their capabilities. This necessity has to be paid additional attention when entering into cooperative arrangements. Cooperation with others could be profitable to all parties but leave some off better than others; that is to say, while cooperation might entail benefits for all in absolute terms, in comparison, some states' shares of the gains might be disproportionately higher than others. This in turn would mean that in relative terms state A, which has benefited from cooperation but not as much as state B has, would find itself in a less favorable position vis-à-vis state B than was the case prior to their cooperation. As states do not know how the state of relations will be in the future, they must worry that they might have contributed to the strengthening of a future foe. This is why, in realist thinking, states worry much more about relative gains and why others' shares are a constant factor in their own calculations.

Another constant in states' thinking is the fear of being cheated upon. There is no guarantee that the partner will not renege on the agreement or reap the benefits and defect thereafter. What if you fulfill the requirements of the agreement and the other party does not? In most of the cases, there will be nobody to force the defecting party to correct the injustice done. As Jervis puts it (1978, 167), in an anarchic world, cooperation that might entail benefits for all parties involved can bring disaster if others defect. Thus, states will be inclined to think twice before entering into cooperative arrangements with other states and might prefer not to do so, even in the face of anticipated gains for all parties (see also Grieco 1993a, 119).

Nonetheless, states sometimes choose to cooperate despite persistent concerns over relative gains and defection. For example, cooperation can occur as a consequence of the balance-of-power logic; that is to say, states might unite against mighty states or against a more powerful coalition in order to balance their power and guarantee their own survival. However, cooperation, in the words of Mearsheimer, is "difficult to achieve and always difficult to sustain" (1994, 12). At the same time, cooperation is not restricted to *friendly relations* and occurs between foes and friends alike. What is of importance is that the arrangements made "roughly reflect the distribution of power and satisfy concerns about cheating" (Mearsheimer 1994, 13). A case in point would be an arms agreement between rivals.

Despite the fact that institutionalists, in comparison to realists, are more optimistic about the likelihood of cooperation and the effects international institutions have on interstate relations, the picture they have of world politics does not seem to be much different from that of the realists. They do acknowledge that states are the main actors (Keohane 1989, 1). The system is anarchic and there is a high degree of uncertainty (Keohane 1993, 332). Institutionalists do also not negate power politics and admit that institutions "are rooted in the realities of power and interest" (Keohane & Martin 1995, 42), implying that institutional effects will be dependent on structural constraints. It is the distribution of power among states that determines the "incentives and prospects for international regimes" (Stein 1993, 48).[34] For cooperation to occur and institutions or international regimes to be established - i.e. "sets of implicit or explicit principles, norms, rules, and decision-making procedures around which actors' expectations converge in a given area of international relations" (Krasner 1982, 186) - states have to realize certain mutual interests and opportunities for mutual gains (Keohane 1989, 2).[35] Moreover, there may be different solutions to an existing problem and states may face the necessity to decide on a particular kind of institution. This decision will be shaped by the existing power realities (Keck 1997b, 265). Nevertheless, institutions are established by consensus, which constrains power politics even if it cannot be negated as a whole. Apart from the issue of power politics, institutionalists, just as the realists, also identify distributional problems and fears of defection as main factors impeding cooperation. However, they believe that realists overemphasize conflict and underestimate prospects for cooperation (see Grieco 1993a, 117). Notwithstanding the fact that agreements are neither made nor kept with ease, in an interdependent world there is and will always be a need for collaboration and coordination if states want to achieve gains unattainable on their own, solve problems, or avoid certain outcomes. This need for cooperation may, in certain cases, necessitate the creation of international institutions which are expected to mitigate the effects of anarchy and play a much more important role than realists admit (Keohane 1989, 2).

---

[34] A study by Duffield (1992) on NATO conventional force levels in the central region during and in the aftermath of the Cold War serves as an example to understand the interplay of regime and structural factors. According to Duffield, the fact that conventional force levels in the central region remained relatively constant during the 1960s, 1970s, and 1980s, despite variations in the extent of the actual or perceived Soviet threat and changing budgetary situations, can be attributed to the existence of a regime which facilitated an adherence to the status-quo and made unilateral actions costly (Duffield 1992). Duffield draws attention to the fact that this regime ceased to exist after the demise of the Soviet Union and suggests that underlying changes in structural conditions will result in the replacement, abandonment, or the modification of a regime, whereas modest shifts in the distribution of power need not necessarily have such consequences. He draws the conclusion that one "may regard structural factors as setting only broad limits on international outcomes, including the possible forms that regimes may take.[...] A variety of regime rules may be consistent with a given distribution of power. Once a regime arises, however, it further shapes and constrains state behavior within the parameters established by the structure of the system. Thus, balance-of-power theory and the public goods theory account for the range of NATO conventional force levels that were possible, while regime theory explains why the allies adhered to a particular set of force levels within that range. In this sense, structural and institutional approaches to understanding international relations are highly complementary" (Duffield 1992, 854).

[35] Hegemonic power can also contribute to the construction of international institutions, but it is not indispensable for their continued existence if mutual interests necessitate further cooperation (Keohane 1984, 12).

While institutional mechanisms can facilitate cooperation in a number of ways, the reader will only be presented with a brief summary, and only those functions which seem to be of relevance with regard to the topic of this work will be singled out (for a detailed coverage of functions institutions could perform see, for example, Martin & Simmons 1998, 745-746; Keohane 1974, 244-245; Keohane 1989, 12-13, 166-167; Keohane & Martin 1995, 42; Mearsheimer 194/95, 16-18; Grieco 1993a, 123-124). Institutions, first and foremost, provide forums for consultations and communication and facilitate the flow of valuable information. Such information is of outmost importance when trying to ally partners' concerns about one's motivations and intentions and for gaining insight into the capabilities, motivations, and expectations of other member-states alike. The fact that, institutions allow not only for formal forums but also for informal contacts and communication among officials renders them even more important for states (cf. Keohane 1982, 349). These contacts are expected to give rise to trans-governmental networks of acquaintance and friendship. Officials will be involved in informal discussions, elaborate policies and exchange valuable (sometimes even secret) information. Consequently, like-minded officials might join forces to achieve common purposes.[36] On the whole, such contacts will facilitate cooperation because they provide governments with valuable information about what the other states intend to do.

As a further consequence, the flow of information is believed to enable states to identify attempts to defect at an early stage and undertake countermeasures. If cheating nevertheless occurs, who the defectors are will be visible. Having identified defectors, punitive action can be taken against such states. What makes punishment possible is the fact that institutions allow for the iteration of the *game* and enable states to pursue tit-for-tat strategies (see also Lipson 1993, 63-65). Tit-for-tat strategies refer to the reciprocation of action among states. You can expect rewards only for cooperative behavior. If you defect at a certain stage of the cooperative arrangement, you must know that you will be punished for your behavior as others will defect in following stages. The iteration of the game within the framework of institutions forces states to take into account the future as well (see also Keohane 1984, 76-77)[37]. Thus the shadow of the future looms over the

---

[36] Jervis points to the example of the Concert of Europe where representatives to the conferences met often and spent a long time together, which helped them adopt a "common outlook distinct from their governments" (1982, 367).

[37] Such reasoning might have played a role during the Cuban missile crisis and the Berlin crisis (Haftendorn 1997,15). Haftendorn argues that in such crisis situations the NATO allies displayed a great degree of solidarity and willingness to cooperate. Their readiness to cooperate might have been conditioned by the wisdom that states that defected now could not count on the assistance of others during a future crisis situation. Another factor might have been the fact that the allies

states and influences their calculations. Especially when cooperation proves profitable and a state is keen on further gains in the future, it will tend not to endanger these prospective gains by defecting. Thus such mechanisms are likely to discourage cheating, as defectors know from the very beginning that they will be caught, punished, and forced to forgo future gains.

In the meantime, the expectation that the norm of reciprocity underlies exchanges between the members of a regime makes it easier for states to make concessions (cf. Jervis 1982, 367). States do not have to worry that their concessions will be interpreted as a sign of weakness inducing the counterparts to demand more concessions. Concessions are simply regarded as a rule of the game and are expected to be reciprocated.[38] The reciprocation of action is facilitated by the opportunities for issue-linkages and side-payments institutions provide. Concessions made in one field can be reciprocated in another issue-area. In the same manner, defections in one issue-area can be punished by defecting in other issue-areas. Keohane refers to this kind of linkages as *retaliatory linkages* (Keohane 1984, 104).

Furthermore, international organizations can act as mediators and prove to be of help in resolving disputes (Abbott & Snidal 1998, see also Russett/Oneal/Davis 1998, 444). They can do so by acting as an "honest broker" (reducing transaction costs, improving communication and the flow of information between the disputants, especially about their preferences, overcoming bargaining deadlocks). Abbott and Snidal refer to such kind of services as "facilitative intervention" (Abbott & Snidal 1998, 22). A point in case would be the good-offices provided by the Secretary General of the United Nations. Beyond this, some institutions can issue legally binding decisions. Such "binding interventionism" (Ibid., 23) can, on the one hand, provide an incentive against rigid bargaining positions, since states would have to expect foreign intervention whose outcome they would not be able to anticipate if bilateral negotiations failed. On the other hand, binding decisions of an international authority can be sold to the public more easily than

appreciated NATO not only for the purpose of collective defense, but also for other reasons. Therefore, they might have worked together to secure the continued existence of the Alliance.

[38] Granting concessions to others becomes less difficult for states, too, because institutions serve as face-saving forums. Concessions made within the framework of international institutions can be more easily sold to the public which usually displays greater misgiving towards concessions made vis-à-vis other states. In this vein, Abbott and Snidal maintain that international organizations make activities possible that would not be possible on a bilateral level (1998, 18). For example, accepting financial aid from international monetary institutions like the International Monetary Fund (IMF) might be more favorable to states than borrowing from the United States. Domestic arguments of surrendering the independence of the country or arguments of inferiority would constrain decision-makers much more when borrowing from a great power on a bilateral basis.

54

compromises made vis-à-vis another state. For example, the European Court of Justice is ensued with the authority to act as a final arbiter in disputes among EU member-states or between the commission and a member-state. Apart from acting as a mediator, international institutions can act as trustees or allocate (scarce) resources (Ibid., 22).

What is more, institutions could also stimulate changes in the domestic politics or the political culture of states, which in turn might entail a redefinition of interests and preferences (cf. Haftendorn 1997, 22-23).[39] Haftendorn points to the importance of common values and refers to NATO as a case in point.[40] International institutions, she argues, promote the parallelism of states' interests and values. Officials or even the public may internalize the norms of an institution (McCalla 1996, 262). Similar or common values in turn make the behavior of states more predictable and encourage states towards further cooperative behavior. Consequently, common values are expected to have a positive effect on the cohesion and performance of institutions.[41]

---

[39] The case of Germany is often cited as a case in point. According to this generally offered view, Germany was enmeshed in a web of international institutions in order to be pacified, democratized and bound to the West. For example, Ingo Peters draws attention to the fact that Germany has, in contradiction to realist predictions, according to which it should have more and more preferred unilateralism to multilateralism with the Cold War gone, continued its tradition of reliance on multilateral policies (1999, 196). He argues that Germany's embeddedness in a web of international institutions turned multilateralism into a "causal belief" for the Germans – that is, it is "taken to be an indispensable prerequisite for achieving one's preferences" (Ibid., 198). This embeddedness and related causal beliefs are, in Peters' view, responsible for German policies toward the OSCE in the post-Cold War era. Multilateralism is no longer a policy-means for Germany but much more an end in itself, so Peters (Ibid., 201). In the meantime though, German commitment to the OSCE served to gain the approval of the Soviet Union for German reunification and to meet related concerns of other states (Ibid., 203, 216). Peters concludes that German preferences remained constant despite changes in the international system due to its membership in international institutions (Ibid., 217).

[40] Even though Haftendorn refers to NATO as a value-community and points to the positive example of Germany, when it comes to the claim made by the proponents of NATO enlargement according to which such a move would promote democracy in the prospective member-states and shape their domestic politics making their foreign policy behavior more predictable, she identifies a necessity to examine whether or how far the changes can be traced back to the direct or indirect effects of institutions, or whether other factors, like a redistribution of power among states, changes in other regions, or significant domestic developments have played a greater role (1997, 26).

[41] Russett, Oneal and Davis argue that collective decisions are legitimized by prior discussions, which in turn promotes adherence to the decisions taken (1998, 446). Furthermore, common norms are expected to bring about common interests and therefore facilitate cooperation.

Peace via International Institutions?

The central question remains whether there exists a direct link between institutionalized cooperation and international stability and peace. Even though realists like Mearsheimer acknowledge that states sometime prefer to cooperate, they adopt a pessimistic view towards the merits of international institutions. Mearsheimer rejects a causal link between cooperation within the framework of institutions and peace (1994, 7, 13-14). In his view, institutions do not have an independent effect on state behavior and merely reflect the distribution of power. They are established by powerful states aiming at managing, maintaining, and expanding their power positions. Consequently, it was the balance of power between the two blocs that was responsible for the long peace during the Cold War era, not NATO *per se*.

Institutionalists, on the other hand, hold that institutions, providing aforementioned services to self-interested nation-states, could affect the interests and preferences of states and consequently have an impact on their behavior. Accordingly, Keohane argues that institutions, altering the payoff structures and providing insight into the calculations of others, affect the incentives facing states, which in turn shapes their behavior (Keohane 1989, 5-6). Yet, the claim made by instititutionalists is not that international institutions will always be of significance or unfold the same positive effects under all conditions. Institutions are subject to the confines of power and interest and their effects will vary in different settings (see Keohane & Martin 1995, 42).[42] Thus Keohane and Martin explicitly warn against exaggerated expectations and note (1995, 50):

---

[42] In line with these assumptions, Martin and Simmons formulate a typology of institutional effects and state that, "institutions, or perhaps similar institutions in different settings, will have different types of effects" (Martin & Simmons 1998, 752). They differentiate between convergence and divergence effects of institutions (Ibid., 752-757).

For a typology of institutions see also Jervis 1999, 55-62: The main criteria making the difference, according to Jervis, is whether these institutions solely facilitate cooperation by easing concerns over defection and unbalanced gains, or beyond that, whether they change preferences over outcomes. Accordingly, he differentiates between (a) *institutions as standard tools* - institutions which embody state interests and fulfill the task of binding states and their partners to deals struck (e.g. alliances, trade agreements); (b) *institutions as innovative tools* - states realize certain new areas of common interest where institutional mechanisms could foster cooperation, or where better use could be made of existing under-utilized institutions; (c) *institutions as causes of changes in preferences over outcomes* - such institutions cause unanticipated and unintended effects and entail a change in the preferences of the member-states.

Another point to be made here is that institutionalists assert that the solution of different cooperation problems will require different institutional features (cf. Haftendorn, Keohane & Wallander 1999, 7-8). They differentiate between collaboration, coordination, assurance and suasion situations (see also Martin 1992, 768-783; also Riecke 1997).

> The necessity for institutions does not mean that they are always valuable, much less that they operate without respect to power and interests, constitute a panacea for violent conflict, or always reduce the likelihood of war. Claiming too much for international institutions would indeed be a "false promise." But in a world politics constrained by state power and divergent interests, and unlikely to experience effective hierarchical governance, international institutions operating on the basis of reciprocity will be components of any lasting peace.

What seems to be of importance is the *assumption of conditionality* in institutionalist theory. Institutionalists reject the view that neorealism and institutionalism are "diametrically opposed theories" (cf. Keohane 1993, 277-278) and argue that one has to pay attention to the underlying conditions under which one or the other approach is valid. According to Keohane, in situations where states share mutual interest and expect cooperation to entail substantial gains, neorealism's evaluation of the prospects for cooperation will diverge from neoliberal institutionalist predictions. However, when states do not share substantial common interests and are deeply concerned about relative gains, both theories' expectations will converge (see also Keohane 1989, 14-16). In a later work, Keohane, together with Haftendorn and Wallander, holds the view that institutionalism, on the one hand, incorporates much of realism, but on the other hand, supersedes realism, "offering a richer specification of the institutional and informational environments within which strategic interaction takes place" (1999, 335).

The affinity between neorealism and neoliberal institutionalism is also acknowledged by Robert Jervis, who claims that differences between both lines of thinking have been exaggerated and misunderstood (cf. 1999, 47). According to Jervis, representatives of the respective schools differ on the question of how much conflict is unnecessary and avoidable. Neoliberals, in contrast to realists, tend to see more opportunities for states to cooperate than they realize. Jervis identifies a necessity to differentiate between offensive and defensive realists (cf. Ibid., 48-50). Offensive realists think that many states might be willing to go to war to expand or may have "security requirements that are incompatible with those of others" (Ibid., 48). Jervis cites John Mearsheimer as a representative of this branch. Defensive realists, on the other hand, are closer to neoliberal institutionalist thinking, since they support the view that conflict is only unavoidable in cases where states have aggressive designs or where states' interests are irreconcilable. However, in most of the cases conflict arises as an unintended result of states' efforts to guarantee their security given the effects of anarchy, especially in the face of the security dilemma. In the terms of Jervis, "international politics represents tragedy rather than evil as the actions of states make it even harder for them to be secure" (Ibid., 49). Even though Jervis maintains that there are still differences between the conceptions of defensive realists and neoliberals - the most important being the fact that defensive realists are less optimistic about the prospects for states to overcome mutual mistrust and fears of cheating in order to cooperate - he points out that defensive realists are sometimes closer to neoliberals than to offensive realists. This is the case when status quo states face like-minded states (Ibid., 52):

> When dealing with aggressors, increasing cooperation is beyond reach, and the analysis and preferred policies of defensive realists differ little from those of offensive realists; when the security dilemma is the problem, either or both sides can seek changes in preferences over strategies (both their own and those of the other) in the form of implementing standard "cooperation under anarchy" policies. In these cases, defensive realists and neoliberals see similar ways to reduce conflict. Both embrace the apparent paradox that actors can be well advised to reduce their own ability to take advantage of others now and in the future. Both agree that cooperation is more likely or can be made so if large transactions can be divided up into a series of smaller ones, if transparency can be increased, if both the gains from cheating and the costs of being cheated on are relatively low, if mutual cooperation is or can be made much more advantageous than mutual defection, and if each side employs strategies of reciprocity and believes that the interactions will continue over a long period of time.

Returning to the *concept of conditionality*, the assumption that "institutions have an interactive effect, meaning that their impacts on outcomes varies depending on the nature of power and interests" (Keohan & Martin 1995, 42), renders it necessary to see the effects of institutional *attributes* such as transparency, issue linkage, or consultations and communication - all expected to contribute to cooperation and stability - in perspective.

To begin with, issue-linkages can be generally employed by negotiating parties in a constructive manner for the purpose of facilitating cooperation, just as they might be used for forcing others into agreements they would otherwise not accept.[43] Beyond that, adding issues where no agreement seems possible to

---

[43] Joanne Gowa agrees that the effects of linkage are conditional (1989, 319): "As both theoretical and empirical analyses suggest, linkage may as easily torpedo as reinforce cooperation in any

negotiations about less contentious issues and demanding joint resolution might preclude chances for a deal on the less divisive issues as well (Sebenius 1993, 300). In such cases, it might be better to "subtract" these contentious issues in order to enable agreement on other issues. In a similar fashion, Axelrod and Keohane hold that issue-linkages "have dangers as well as opportunities" (Axelrod & Keohane 1993, 100). For example, issue-linkages can prove destructive when a state demands too much in an added issue-area, making agreement in the area where mutual interests exist impossible. In general, where no shared interests exist, issue-linkages are not expected to be of help and could even impede cooperation (Keohane 1993, 278). Thus regimes are not only expected to facilitate issue-linkages, but also to prevent those linkages which might render cooperation more difficult (cf. Keohane 1984, 91). Keohane refers to successful regimes as those that facilitate *productive linkages* - i.e. "those that facilitate agreements consistent with the principles of the regime" (Ibid., 92) - and simultaneously discourage *destructive linkages* that are in conflict with regime principles. In a similar fashion, Hasenclever asserts that one way in which *interdemocratic institutions* can foster cooperation is by preventing conflict from spreading from one issue-area to other issue-areas (2002, 98-999). Interdemocratic institutions enhance the autonomy of single policy areas and allow for the existence of what Hasenclever labels *islands of cooperation* enabling states to continue cooperation in one field in spite of conflicts in other areas (Hasenclever 2002, 90). Thus, in situations where two countries are in conflict with each other, the task for *successful* institutions would be to limit conflict to certain areas and prevent its extension into other issue-areas. In this connection, Wallander asserts, "NATO's structure ensures that progress becoming stalled in one area will not prevent progress in another" (quoted from Hasenclever 2002, 99).

As is the case with issue-linkage, the institutionalist argument on the conditionality of institutional effects can be viewed as applicable to transparency as well. In situations where a state's articulated intentions are not benign, transparency cannot be expected to facilitate cooperation. Nor can states ever be absolutely sure that the information provided by their partners, be it on capabilities or on intentions, will be reliable with a hundred percent certainty. The degree to which such information will be reassuring will be dependent on the nature of the relationship at hand. Foes will be, if at all, more difficult to reassure than friends. Even if the information was reliable, the possibility that capabilities as well as intentions might change in the future is always given. Therefore, no

specific issue area: the interests of states in linking cooperation on one to cooperation on other issues can as easily diverge as converge. Analytically, then, there is no reason to assume that linkages stabilize cooperation."

matter how well an institution works and how sincere states' pronouncements are, a certain amount of *residual* uncertainty and insecurity will always remain. This does not conflict with institutionalist argumentation, since institutions are expected to mitigate the effects of the security-dilemma, not to neutralize it as a whole.

Apart from the fact that issue-linkage and transparency need not necessarily unfold positive effects and could even worsen the state of relations, Ronald Krebs argues in his essay on the same topic that membership in the same alliance can also deepen and intensify conflict among feuding members due to arms transfers by big powers to smaller states or due to the desire of the feuding members to *capture* the institution in order to be able to further their position vis-à-vis the opponent (cf. Krebs 1999)[44]. In Krebs' view, under normal circumstances, all

---

[44] According to Krebs, under certain circumstances, membership in the same alliance might also lead to the revival of conflict between two small powers (cf. Krebs 1999, 345, 351). Small powers, having been admitted into an alliance and having gained the security guarantee of powerful allies against a common security threat which they would otherwise not be able to counter, may experience a shift in the assessment of their foreign policy objectives. So they may be inclined to focus on their antagonisms with an allied long-time rival since the burden of dealing with the major threat has been transferred to other powerful alliance members. In the view of Krebs, accession to NATO had such an effect on the relations between Greece and Turkey. A 20-year-period of détente was to end in the mid-1950s when Greece intensified its efforts for *enosis*. Prior to admission to NATO, both Greece and Turkey had to deal previously with Bulgarian and Italian expansionism and later with the Soviet threat on their own and could not pay attention to such "petty" and "peripheral" - as Krebs puts it - regional issues like Cyprus. Having joined the Alliance, however, Greece and Turkey were able to refer the task of defense against possible Soviet bloc aggression to their major alliance partners and could now turn their attention to regional objectives which had earlier been suppressed; that is to say both countries were seduced into reassessing their foreign policy objectives due to the guarantee provided by NATO.

While this thesis will not be dealt with in more detail here, it is worth mentioning that an analysis of the factors responsible for the period of détente and its end in the mid-1950s, which was conducted within the framework of my masters thesis submitted at the University of Vienna, showed that factors unrelated to their admission to NATO conditioned their behavior. While the security-guarantee provided by NATO might have played, at best, a minor role in the calculations of Greece and Turkey, it did not condition a foreign policy shift as asserted by Krebs. The main factor putting an end to the era of détente had been the drive for enosis which had gained momentum in the early 1950s due to various reasons, including the fact that the British had transferred the *responsibility* for the defense of Greece and Turkey against communism to the United States, that some degree of domestic consolidation and stability had returned to both countries by the mid-1950s, that the enosis movement on Cyprus had gained momentum in the early 1950s with the holding of a plebiscite and the appearance of Makarios III as Archbishop - who was to appeal to the UN for the first time in 1953 on his own - and due to the fact that governments in both countries came under pressure by the public to assert responsibility for the fate of their compatriots.

allies should welcome increases in the capabilities of other alliance members as such increases are expected to add to the security of the alliance as a whole (cf. Ibid., 152-154). The problem arises when the intra-alliance security dilemma starts to operate and feuding allies become worried that these newly acquired capabilities could one day be used contrarily to their original purpose against other alliance members. Thus arms transfers providing additional capabilities feed mutual suspicions and cause heightened tension between allies. The parties turn out to be less willing to make concessions in order to display resolve, rendering cooperation on issues burdening their relations more difficult. Maintaining that offensive and defensive weapons cannot be distinguished, Krebs draws the conclusion that "military assistance of whatever nature seems likely to contribute to the deterioration of relations between the coalition's allied adversaries,

---

Moreover, no evidence can be drawn from the cases of Greece and Turkey to support the claim made by Krebs according to which the fear of abandonment might work against a redirection of interest to bilateral disputes with neighbors. Greece and Turkey have had such uncertainties related to the reliability of the Alliance for decades. There had never been a 100 percent certainty that NATO would really assist these countries in the case of a Soviet attack. The withdrawal of the US Jupiter missiles during the Cuban missile crisis without prior consultations with Turkey, the wording of the 1964 Johnson Letter saying Turkey could not count on assistance in case of a Soviet involvement triggered by a Turkish military intervention on Cyprus, or the US arms embargo imposed on Turkey after the invasion of Cyprus can all be viewed as clear cases where abandonment was practiced or at least threatened. The credibility of the NATO commitment was even questioned by former SACEUR General Rogers who, according to a report published in the Turkish daily *Hürriyet* on October 25th, 1987, admitted he had doubts whether some NATO allies, for example the Northern Europeans, would come to the aid of Turkey if the Soviets entered Anatolia (cf. Bulaç 2001, 68). Meanwhile, reading through the works of Greek authors, one has the impression that the uncertainty felt in Greece with regard to the credibility of the NATO commitment was by no means a lesser one. For example, Iatrides claims that the security guarantee provided by the Alliance held more of a symbolic value for Greece, and "in the calculations of NATO and American strategists, Greece was from the onset a peripheral partner, perhaps even an expendable one" (1993, 14). He adds that NATO was not willing to invest the necessary resources in the defense of Greece. NATO plans could not guarantee that an aggression against Greece would be repelled. In his view, Greece was to fulfill the function of a "trip-wire". If Greece was attacked, the Alliance would recognize it was at war, "but in fulfilling its important purpose a trip-wire is destroyed" (Iatrides 1993, 14). In a similar fashion, Veremis maintains that the Greek forces were expected to cause some delay to invading Warsaw Pact forces in case of war but had not been equipped adequately to fulfill such a task (1984, 33). Note also that certain features that work against the problems of abandonment and entrapment exist within the framework of NATO. While Article V was formulated in a manner that would exclude automatic military involvement, a measure which was necessary to gain the support of the US Senate for the Treaty (Interviews in Brussels, July 2001) and thus mitigates fears of entrapment, the military planning of the Alliance simultaneously works to mitigate the fears of abandonment. Overall, NATO membership did not take the burden of defense from their shoulders as a whole, and the possibility of abandonment has always been given in the Greek-Turkish case, but neither prevented the revival of conflict, nor hindered its continuation.

assisting in the process converting a limited dispute into a broad security threat" (Ibid., 354). At the same time, states might try to raise the contested issues in institutional forums and try to generate support for their own position and manipulate the distribution of benefits to their ends (cf. Ibid., 355-356). As a consequence, the institution itself might become the object of desire and add to the list of contested issues.[45]

International Institutions and Democratization

Ever since the Cold War ended, proponents of NATO expansion have asserted that an inclusion of former Soviet bloc states in NATO would promote democratization in the region. As it was stated in the *Study on NATO Enlargement* (1995),[46] NATO, too, identified expansion as a mechanism for "[e]ncouraging and supporting democratic reforms, including civilian and democratic control over the military". NATO's former Secretary-General Willy Claes made it clear that aspirant countries could not expect to be admitted without having persuaded NATO of their democratic character and their respect for human rights (Pourchot 1997, 159).[47] Democracy is not only favored as an end itself, but also as a means for enhancing security and stability in the region given the expectation that democracies do not fight each other (see Reiter 2001, 41).

---

[45] While Krebs identifies a necessity to moderate institutionalist claims and develops what he calls "realist institutionalism" which rests on the assumption that institutional effects on inter-state relationships need not necessarily be positive and that alliances can even intensify conflict among feuding members (cf.1999, 344-345), the assertion made here is that this core wisdom of realist institutionalism is already inherent in neoliberal institutionalism and does not constitute a novelty. As shown above, institutionalists are aware of the fact that transparency or issue-linkage cannot have stabilizing effects under all conditions. The same wisdom applies to the concept of institutional capture brought in by Krebs. Institutions can be of value to states for many reasons. These reasons need not always be benign, virtuous, or moral. Institutions are more often than less regarded as tools - tools that can be employed to achieve ends which might be unrelated to the original purpose or even be incompatible with the underlying norms and principles. The world has often had to witness how UN forums have been used for legitimizing war rather than for preventing its occurrence.
The fact that institutions could be employed for *less divine* purposes is not alien to institutionalists. It constitutes a wisdom they would not challenge. As stated above, neoliberals conceive institutions as acting within the confines of state interests and power. If a state has an interest in preventing an allied adversary achieving ends in conflict with its own interests, it would not be of surprise to institutionalists if this state tried to employ the institution for these purposes.
[46] The Study on NATO Enlargement can be viewed on of NATO's website (www.nato.int/docu/basistext/enl-9502.htm).
[47] Further criteria Claes mentioned were interoperability of arms and "guarantees that it will not import regional crisis into the Alliance" (Pourchot 1997, 159).

The main theoretical question to be answered is whether institutions can have an impact on the internal characteristics of states and act as catalysts for internal transformation processes. As stated above, neoliberal institutionalists assume such a relationship when they refer to socialization processes within institutional frameworks and assert that institutions can alter the preferences of member-states. Unfortunately, they do not further elaborate the issue or provide an explanation of underlying mechanisms. An attempt at providing a theoretical framework for "outside-in" linkages, that is to say, the causal links between intergovernmental organizations and states, though is undertaken by Jon C. Pevehouse, who finds that such linkages do play an important role (Pevehouse 2001).

Pevehouse identifies a number of mechanisms allowing external factors to shape internal transformation processes. First, autocratic regimes facing a crisis (e.g. economic crisis or political uprisings) might see their standing jeopardized and feel the necessity to liberalize in order to re-legitimize their grip on power (cf. Pevehouse 2001, 521-524). Once such a liberalization process is started, intergovernmental organizations (IOs) can adopt economic and diplomatic measures or even expel the regime, which might push the liberalization process much further than would otherwise be the case.[48] Economic sanctions might work to further erode the regime's economic basis and destabilize it, while diplomatic pressure could enhance its isolation and further de-legitimize its reign at home and abroad. In the end, the regime might give in to the pressures and democratize.[49]

---

[48] IOs can be preferred to unilateral efforts because their multilateral nature will make the involvement in a state's internal affairs appear less offensive and offer opportunities to member-states to express their opinions (Ibid., 523). In general, compared to bilateral exchanges "multilateral allegiances usually prove more benign for democratization" (Pridham 2000, 292). Compared to institutional frameworks, pressure exerted by another state unilaterally could be perceived as illegitimate offences against its sovereignty and could thus backfire and generate support for the authoritarian regime (Pevehouse 2001, 523). Pevehouse cites the pressure exerted by the EC on the Colonels' regime in Greece as a successful case of institutional interference (2001, 524). The EC is said to have contributed to the weakening of the Colonels' regime, since the Association Agreement between the EC and Greece was suspended after the coup d'état in 1967 entailing serious economic losses for Greece and isolating it from the integration process.

[49] Democratic states might choose to force others to democratize for quite selfish reasons. Pevehouse maintains that democracies might push for democratization in other states in a move to further their own internal and external legitimacy (2001, 522).Such a behavior is expected to nourish national pride and self-confidence. Pevehouse cites US efforts for democratization as an example (2001, 523).

However, even though such a motivation might be true for some states, especially weak ones which might not have any strategic interests involved, for global players like the US pursuing a multifaceted set of interests based on and aimed at securing and advancing its economic and military hegemony the "promotion of democracy thesis" will first and foremost serve as a pretext for intervention and as a justification thereof at home and abroad.

Secondly, IOs can contribute to liberalization processes by assuaging the fears of the elites of democracy either by providing certain safeguards protecting their rights and interests or altering their belief systems (Pevehouse 2001, 524-529). Especially business elites and the military might be worried that a regime change could undermine their position and run counter to their interests, thus causing them adopt a stance inimical to democratic transition. IOs' binding function might be of great value in this context. Since IOs require from states their continuous commitment to deals struck and agreements made, even in the face of government or regime changes, business elites might see their interests as less threatened in the face of liberalization processes and acquiesce to democratization. For example, the guarantees given by the EC to business elites in Southern Europe are said to have played an important role in the transition processes in the region (Pevehouse 2001, 526-527). Any properties taken by the state had to be compensated for and the free movement of capital and goods were guaranteed. In the view of Pevehouse, these guarantees provided by the EC were especially important in the cases of Spain[50] and Portugal, "where economic elites had traditionally been hostile to democracy" (Pevehouse 2001, 527). With regard to the military, IOs can help to win over their acceptance of democratization by providing additional resources and meeting their material demands. Hungary, for example, received assistance in its modernization efforts from NATO and was asked to increase military spending. Pevehouse asserts that NATO membership was important for the transition process because NATO policies and assistance helped to satisfy the demands of the military which had recently been feeling uneasy given the decreasing budget levels.

IOs also allow for contacts between military officers from different member-countries. Officers from young democracies or autocratic countries become familiar with democratic forms of civil-military relations and are expected to internalize the idea of civilian control over the military. Or to put it in other words, they accept democratization as a result of the socialization process within

---

Another reason for states to push other states to democratize might be findings showing that democracies "prefer to trade, cooperate and ally with one another. In addition, democracies better promote economic stability and growth (Pevehouse 2001, 523)".

[50] During the Franco regime, the Spanish armed forces' main task had been to prevent domestic unrest (cf. Kay 1998, 56). Because Franco had established what Kay calls an "officer-heavy patronage system" to secure the loyalty of the armed forces, the military resisted attempts to democratize in the late 1970s. Thus one of the reasons why Spain wanted to join European institutions like NATO was the belief that this would facilitate the democratization process and help to avoid any other military intervention (Kay 1998, 56).

Yet, Spain's desire to join NATO was also conditioned by the hope that membership in NATO would render it less difficult to join the EC (Kay 1998, 57). The US' desire to have Spanish territory included in NATO defense planning also contributed to its admission to NATO.

the framework of an IO. Pevehouse cites the experience made with the Spanish military as a case in point (2001, 528-529). In his view, NATO membership proved essential in modernizing the Spanish armed forces and re-directing their attention away from domestic politics to external functions.[51]

Despite the fact that international institutions may make a difference with regard to the prospects for democratization, Pevehouse adds that not all IOs can have the same positive effects on democratization (2001, 529-531). It is IOs with a high democratic density – that is to say, with a high share of democratic members - which are expected to promote democratization. This is due to a number of reasons. First, democratic IOs are more likely to make membership conditional on fulfilling certain criteria,[52] pressure for their implementation and punish violations. In line with this argumentation, Greece's association agreement with the EC was frozen in the wake of the 1967 coup, and Turkey's involvement in the European Council was suspended after the overthrow of the government by the military in 1980. Secondly, the transparency of democracies makes cheating less likely as cheaters will be detected easily. Therefore member-states are inclined to abide by the rules and contribute to the enforcement of measures against states with undemocratic practices or character. Socialization into democratic practices and thinking is also more likely to occur in IOs with a high democratic density.

---

[51] Kay also maintains that after entering NATO the Spanish forces started to prepare for tasks other than internal control (1998, 57): "The Spanish military assumed five key tasks to complement NATO planning: assure security on the Iberian peninsula; contribute to the strengthening of the defense of the western Mediterranean flank; participate in keeping the Atlantic routes open and assure the aero-naval passage between the US and Spain in the event of conflict; monitor and control the approaches to the Strait of Gibraltar; and integrate the Spanish air-warning network into the NATO-wide early warning systems."

[52] Pridham describes conditionality "as the most suggestive of deliberate efforts to determine from outside the course and outcome of regime change, excepting of course control through foreign occupation" (Pridham 2000, 297). However, he is quick to add that the success of this strategy is dependent on the responsiveness of the country in question.

Conditionality is central to the institutional stability theory as well (cf. Skalnes 1998). According to Skalnes, institutional stability theory basically rests on the assumption that international institutions can promote domestic and, in consequence, international stability by shaping domestic politics and constraints (Ibid., 45).

International institutions serve as instruments to influence the domestic politics of countries seeking membership, since membership is made contingent on reforms in various issue-areas and membership will be granted only after these reforms have been implemented and have taken root. Furthermore, membership will support the position of certain groups while weakening the standing of those hostile to the reforms. In the meantime, international institutions might serve to change the preferences of voters or domestic groups. All this is expected to promote domestic stability (see also Balmaceda 1997, 97-98). Finally, the reforms and the stable environment are expected to come to bear on the foreign policies of the countries concerned.

Given the higher number of interactions with democratic states, "the transmission of values and norms about the democratic process is more likely" (Pevehouse 2001, 530). Finally, democracies are expected to keep to international agreements, since a violation of such provisions would entail domestic opposition and "potential audience costs" (Pevehouse 2001 530). Thus, assurances by IOs with a mainly democratic membership will be more credible than commitments made by IOs of less democratic density.

However, IOs might also choose not to enforce democratic conditions. This might be the case because the costs of doing so may be assessed as being too high or because other strategic interests work against their enforcement (Pevehouse 2001, 530-531). In the face of such conditions, IOs might fall short of promoting democratization.

Pevehouse also provides empirical findings supporting his hypotheses (2001, 531-540). Some of his findings seem to be of interest for the research goals of this work. First of all, Pevehouse finds evidence for his thesis that membership in IOs makes a difference. However, he acknowledges that during transformation processes the effects of IOs need not be the most important ones, and that other factors will probably play a more prominent role. Moreover, the data shows that the magnitude of institutional effects increases along with increasing levels of democratic density. Pevehouse draws the conclusion that the data provides evidence that democratic IOs make a difference, yet adds that, "[t]he argument is *not*, however, that IOs play the strongest role in the prospects for regime transition. Indeed, the results of these analyses indicate [...] that internal factors such as past experience with democracy, previous type of regime, and changes in economic growth rates play a far more substantial role" (2002, 539).

Whereas Pevehouse points to examples where NATO played an important role in facilitating democratic reform, this view is categorically rejected by Dan Reiter who asserts that NATO neither contributed to the democratization of member-states during the Cold War, nor promoted democracy in the three former Soviet bloc countries admitted in 1999 – the Czech Republic, Hungary and Poland - and will not play a role in the democratization of the applicant countries (Reiter 2001). According to Reiter, it is primarily economic factors that determine the prospects for the emergence or collapse of democratic regimes and there is little that security institutions can do (2001, 56). Neither the prospect of NATO membership or the enforcement or pressuring by NATO, nor socialization or contacts among military personnel from different member-states have worked to promote democratization, nor will they do so in the future. With regard to the member-states that joined NATO in 1999, Reiter maintains that the democratization processes had begun before the prospect of NATO membership even emerged. These countries had opted for democracy before they opted for NATO (Ibid., 60).[53] Furthermore, the Czech Republic, Hungary and Poland had all implemented measures aiming at guaranteeing the civilian control of the military in the early years of the 1990s, that is to say, long before their membership in NATO became an issue (Ibid., 62-63). Moreover, while NATO policies may have contributed to the strengthening of the civilian control over the military, in fact, no country had really been confronted with the prospect of a military intervention. NATO and PfP might even have certain adverse effects on democratization in these countries. On the one hand, new members are required to upgrade their military arsenals in order to meet NATO standards, a requirement that creates an additional burden on their economies.[54] This, in the view of Reiter, could negatively affect democratization, "given that economic prosperity is one of the most important factors driving successful democratization" (Ibid., 51).[55] On the other hand, PfP measures focus on advancing military expertise, neglecting the civilian dimension. This could undermine the civilian control of the military as officers might conclude that the civilians lack necessary expertise in order to be able to exercise oversight over the military (Ibid., 60-61).

Reiter asserts that NATO was focused on the maintenance of alliance unity rather than on promoting democracy in the member-states during the Cold War (cf. 2001, 57-59). He draws attention to the cases of Portugal, Spain, Greece, and Turkey. Portugal was admitted to NATO as a dictatorship. Spain's transition to democracy and its admission to NATO took place at the same time, and, had Spain not made this transition to democracy, in the view of Reiter, it would still

---

[53] With regard to the fate of democratization in these countries, Michael Mandelbaum claimed that "the [Visegrad] countries under active consideration are precisely those best placed to make a successful transition to democracy and free markets without NATO membership" (quoted in Kay 1998, 106).

[54] One of the institutionalist arguments against enlargement had been the fear that, "by emphasizing the military aspects of reform, the wrong aspects of post-communist transitions may be prioritized in some Central and East European countries" (Kay 1998, 113), that is to say, military reforms might gain precedence over needed economic and political reforms should some states seek to enhance their chances for admission by upgrading their armed forces. For example, in 1996, the Czech Republic, in an attempt to gain the favor of the US and NATO, declared it was willing to purchase 6 F-16 fighter jets (Ibid., 114). According to Kay, the costs of purchasing these jets would have used up 1/5 of the defense budget of the Czech Republic and a total replacement of all MIGs would have required the purchase of 18 more such F-16s.

[55] In contrast to this argument put forward by Reiter, Pevehouse declares support for the argumentation of Przeworski and Limongi who claim that "higher levels of economic development are associated with the continuation of rather than the transition to democracy" (Pevehouse 2001, 537-538).

have been admitted to NATO.[56] The two countries of interest here both experienced military interventions as mentioned above. Turkey faced military coups in 1960, 1972, and 1980, and experienced what was labeled as a "postmodern coup" by the military when the government of Necmettin Erbakan had to resign under pressure from the generals in 1997.[57] The neighbor to the west, Greece, was under the rule of the Colonels from 1967 to 1974. Thus NATO's Cold War record of acting as a promoter of democracy was anything but promising.

Regarding the argument on the socialization of military officers into a democratic culture, Pevehouse and Reiter deliver differing assessments. Whereas Pevehouse finds evidence for the thesis that NATO membership acted to promote the civilian control of the armed forces, Reiter holds that NATO membership brought forward "disparate results" (2001, 58). In contradiction to the argument by Pevehouse that NATO membership helped to modernize the Spanish armed forces and re-direct their attention towards external functions, Reiter draws attention to the fact that some of the officers involved in the coup attempt of 1981 had already enjoyed transnational links with US military officers and were instructed at US military schools. In the end, these links did not have the effect of making democrats out of soldiers. With regard to Turkey, Reiter finds that NATO membership contributed to the overthrow of democracy as well as it contributed to its restoration. On the one hand, contacts with military officials from other member-countries and from NATO did not prevent Turkish generals from intervening in politics and

---

[56] Indeed, even though Spain was not admitted to NATO given Franco's relationship with Hitler during World War II, the US were still interested in close ties (cf. Kay 1998, 56). Franco was valued for his anticommunist credentials and the Spanish territory was of strategic significance given its location at the entrance to the Mediterranean Sea. In exchange for basing rights and the permission to enter Spanish territory without prior consultations in the event of war, the US granted economic aid to the Franco regime and helped to modernize the Spanish armed forces (Ibid., 56).

Nonetheless, leaving aside the question as to whether Spain would have entered NATO even if the transition to democracy had not taken place, one of the reasons why the Spanish elites wanted to join NATO was, as stated in footnote 50, the belief that membership in NATO would facilitate the democratization process and help to avoid another coup by the military (Kay 1998, 56). Yet, given the experiences of Greece and Turkey where the military did not display any hesitation to seize power despite existing ties to NATO, the validity of such logic is questionable.

[57] Güven Erkaya, the Chief of the Turkish Sea Forces at that time, maintains that the military would have carried out a coup if they had not succeeded in countering what he calls the fundamentalist policies of the Erbakan government within the framework of the National Security Council (Baytok 2001, 254). He further quotes another military official as saying, "We have chosen Mesut Yılmaz [the successor of Erbakan as PM] to form the new government. He will act according to our wishes; he will abide by what we are telling him" (translation by the author; for the text in Turkish see Baytok 2001, 258).

overthrowing the civilian government repeatedly, and military coups in Turkey "were taken for what was seen as the good of the country; indeed NATO membership may have contributed to the 1960 coup by raising the expectations of junior officers and inspiring them to rebel against their superiors" (2001, 59). On the other hand, transnational contacts may have been a factor inducing the military to return to the barracks after some time. Similar to the Turkish case, the results of membership in the case of Portugal were also contradictory. Whereas contacts with NATO may have contributed to the overthrow of the autocratic regime by the military, this move in fact contradicted the notion of civilian rule. In addition, leftist sentiments in the Portuguese army were to question the country's NATO membership after the dictatorship had been overthrown (Reiter 2001, 59).

Reiter adds that even if there had been a desire to pressure members to abide by democratic rules or to enforce democratic principles, NATO would not have had needed mechanisms for their implementation at its disposal (Reiter 2001, 53). There are no treaty provisions providing for the ejection of a member-state or the suspension of its membership. Such a provision can, for example, be found in the Charter of the Organization of American States (OAS) which provides the legal basis for the suspension of a state's membership in the event of a military intervention (Ibid., 53). In the case of NATO, neither the Colonels' Greece from 1967-1974, nor the *paşas'* - as the generals are referred to respectfully in the country - Turkey faced any reprisals or had ever to fear abandonment or the suspension of their membership.[58]

---

[58] Apart from the questions dealt with above, there is another dimension of importance to the debate about NATO enlargement and its expected effects on democratization in the countries of the former communist bloc- namely public sentiments about NATO membership and related worries. As Georgeta V. Pourchot puts it, "given that membership will be granted to those countries which can pass the test of democracy, the ensuing question is how responsive elites are to public concerns (1997, 165)." Drawing on existing data on sentiments towards membership, Pourchot shows that, on the whole, the public and the elites in Eastern Europe had differing views on the question of NATO membership. It was especially the ruling elites who favored membership and were willing to accept the burdens associated with membership. For example, the Polish ruling-elite was ready to meet the requirements of NATO membership, that it is to say, of upgrading its forces, by redirecting resources from social programs to the defense budget (Ibid., 166). The public in Eastern Europe, on the other hand, was more concerned about their social and economic well-being and had reservations about the costs of admission. This might have been one reason why the majority of the people preferred admission into the EU to admission into NATO (Ibid., 164). Pourchot maintains that the ruling elites ignored the concerns and preferences of the public and concludes (Ibid., 169), "They [the leading elites] prioritise national defense over social program budgets in a manner reminiscent of authoritarian regimes. Their lack of responsiveness to public needs bodes unfavourably for the internal democratization of the region, lately marked by apathy and disengagement from politics. A piecemeal 'democratization', for elites only, cannot be the Western democracies' long-term goal for Eastern Europe, and would affect the continent's

*****

Having discussed neorealist and neoliberal institutionalist theories at some length and portrayed differing views on the role international institutions can play in democratization processes, it is now possible to focus analysis on those *aspects* or *merits* of membership in the same alliance that could allow for an evaluation of NATO's role in this inter-member dispute. Therefore, in the following sections the task will be:

(1) to present NATO's record of mediation in the Greek-Turkish dispute while paying attention to NATO mechanisms for dealing with internal disputes;

(2) to assess the impact institutional *factors* like issue-linkage, transparency, consultations, and military assistance have had on the course of relations between the two countries in question, simultaneously scrutinizing the concept of institutional capture;

(3) to provide an appraisal of the effects of NATO membership on the course of democratization in both countries.

---

security. If countries in which policy-elites do not attach importance to responsiveness, openness, and consensus-building are included in NATO, the likely results on the overall process of security and democracy in Europe will be negative."

4) NATO AS A MEDIATOR

4.1) Three Wise Men

When the North Atlantic Treaty was drafted, the probability of future inter-alliance disputes was not alien to the participants. During the negotiation process, France had suggested the inclusion of an "article of conciliation" in the treaty (Stearns 1992, 77). However, this proposal had been rejected by other countries on the grounds this would constitute a duplication of existing mechanisms like those of the UN or the ICJ. In 1949, all countries concurred that the main purpose of the Treaty should be collective defense. Nonetheless, the members to the Treaty, taking reference to the provisions of the UN Charter, committed themselves to the preservation of peace and the peaceful settlement of disputes among them.

A short time after the Treaty had been signed, the functioning of the Alliance was hampered by inter-member disputes and the necessity arose to look for ways to deal with these disputes. Particularly the diverging interests related to areas outside the Treaty area led to conflicts among the member-states. France and the US were, for example, at odds because of their competition in Indochina (cf. Kay 1998, 36). In 1956, the Suez Crisis once more showed that the NATO allies, on the one hand, having pledged to come to the aid of one another in the case of aggression from third parties, could, on the other hand, become opponents and adopt conflicting policies and practices elsewhere. Furthermore, the allies failed to respond to external developments in concert; the failure of the NATO countries to formulate a joint response to the repression of the reform movement in Hungary by the Soviet Union in 1956 constituting a case in point (Kay 1998, 37). All these failures made clear that it was necessary to promote consultations among members and to establish mechanisms for dealing with, or, where possible, solving internal disputes.

In 1956, a Committee on Non-Military Cooperation was established which was to be headed by the foreign ministers of Italy, Norway and Canada (cf. Jordan 1979, 38; Kay 1998, 37) - this trio was referred to as the *Three Wise Men*. The committee was asked "to advise the Council on ways and means to improve and extend NATO cooperation in non-military fields and to develop greater unity within the Atlantic Community" (Report of the Committee of Three on Non-Military Cooperation)[59]. The committee urged the member-states to strengthen the political and economic side of NATO and to promote cooperation in the field of

---

[59] The text of the report can be viewed on the NATO website; http://www.nato.int/docu/basictxt/bt-a3.htm

culture, called for extending the scope and depth of consultations and made recommendations for the handling of disputes arising between member-countries. These recommendations were to be approved by the Council, which adopted the "Resolution on the peaceful settlement of disputes and differences between members of the North Atlantic Treaty Organization" on December 13th, 1956. The Council, making reference to the obligation of the parties arising from Article I of the Treaty to settle international conflicts peacefully and reaffirming the commitment of the parties to seek closer economic cooperation and to eliminate conflicting policies in this field, adopted the following provisions related to the settlement of internal disputes:

> *The North Atlantic Council:*
> REAFFIRMS the obligations of all its members under Article I of the Treaty, to settle by peaceful means any dispute between themselves;
>
> DECIDES that any such disputes which have not proved capable of settlement directly be submitted to good offices procedures within the NATO framework before member governments resort to any other international agency except for disputes of a legal character appropriate for submission to a judicial tribunal and those disputes of an economic character for which attempts at settlement might best be made initially in the appropriate specialised economic organization;
>
> RECOGNISES the right and duty of member governments and of the Secretary General to bring to its attention matters which in their opinion may threaten the solidarity or effectiveness of the Alliance;
>
> EMPOWERS the Secretary General to offer his good offices informally at any time to member governments involved in a dispute and with their consent to initiate or facilitate procedures of inquiry, mediation, conciliation, or arbitration;
>
> AUTHORISES the Secretary General where he deems it appropriate for the purpose outlined in the preceding paragraph to use the assistance of not more than three permanent representatives chosen by him in each instance.[60]

---

[60] The resolution can be found in the annex to the report of the committee. I am very much obliged to a NATO official for drawing my attention to the existence of such a report.

On the whole, the resolution establishes the Secretary-General as the locus of NATO's internal conflict management *mechanisms*. However, apart from his right or obligation to bring issues that might affect the interests of the Alliance as a whole to the attention of the Council and to offer his good-offices to the disputing parties in the first step, he lacks any enforcement powers and even his mandate to initiate *informal* procedures of conflict resolution, such as mediation or arbitration, is made contingent on the prior consent of the parties involved. Acquiring such consent is in reality much harder than one might think because the member-governments usually have different views concerning the role to be played by the organization, tend to question the neutrality of the secretary-general, and value his person to varying degrees. Sometimes, the secretary-general might be viewed as inappropriate for such a job due to simple reasons like his nationality. For example, according to Woodhouse, Secretary-General Lord Ismay's offer to mediate between Greece and Turkey was unlikely to be accepted due to the fact that he was British (1982, 71) - Britain thence being a party to the conflict.[61] Moreover, a NATO official doubted whether Lord Robertson would be acceptable as a mediator to Turkey given the fact he had been one of the initiators of the process which was kicked off in St. Malo[62] aiming at establishing capacities for the EU to carry out autonomous military operations, which caused much resentment in Turkey as a non-EU NATO member (Interviews in Brussels, July 2001). Another official doubted whether any secretary-general could be acceptable to Turkey as a mediator, since the post is always held by a European.

Another problem the Secretary-General has to live with is that he has no diplomatic or political representatives and that there are no information gathering

---

[61] The relationship between de Gaulle and Secretary-General Stikker constitutes a further example in point (cf. Jordan 1979, 119). Since Stikker was viewed as entertaining a special relationship with the US and neglecting the views of the Europeans, he was in no position to mediate between de Gaulle and the US when France pressed for a greater say and less American influence.

In general, the office of the secretary-general seems to lack authority. Secretary-generals have to be careful not to offend any member-state or the Council as such. For example, Secretary-General Spaak wanted to play an assertive role and sometimes tried to act as if he was the political leader of the Alliance. However, his understanding of the role of the secretary-general sometimes alienated him from the Council. Jordan maintains that Spaak was isolated from the Council and the member governments towards the end of his tenure (Jordan 1979, 259).

Moreover, the position of the secretary-general as the dominant figure in the organization has at times been challenged by the SACEUR and the US permanent representative (Jordan 1979, 260). When Eisenhower was SACEUR and Lord Ismay Secretary-General, it was Eisenhower who seemed to be the principal political figure in the organization.

[62] The Anglo-French declaration of St. Malo called on the EU to "have the capacity for autonomous action, backed up by credible military forces, the means to decide to use them, and a readiness to do so, in order to respond to international crises" (quoted in Carr & Callan 2002, 110). This position was later embraced by the Cologne Summit of 1999 (Ibid., 112).

services at his disposal putting him in a position where he is forced to rely on the willingness of the member-states to share their information with him (cf. Jordan 1979, 121; Stikker 1965, 7; Stikker 1966, 407).[63] Yet, he may not be able to use publicly information shown to him in confidence (Ibid., 12). In this connection, Stikker notes that when he identified a need to mediate between Greece and Turkey in 1964, he felt the necessity to gather information proving that the Alliance was facing a serious crisis at its southern flank and thus justifying an intervention by the secretary-general (Stikker 1966, 407; Jordan 1979, 135). In fact, the representatives of the US and UK had shown government reports to the Secretary-General according to which the danger of war was imminent. However, the representatives had declined to give him the dispatches because they were confidential (cf. Jordan 1979, 135). As a substitute, the Secretary-General asked SACEUR Lemnitzer whether he could confirm the information put forward by the representatives. Lemnitzer, thereupon, visited Athens and Ankara to draw a picture of the situation. Indeed, Turkey was preparing for an invasion while Greek special units had been put on stand-by and the air, sea, and land forces were conducting joint maneuvers. Thus Lemnitzer could confirm to the Secretary-General that the situation was serious, upon which Stikker could become active (Jordan 1979, 135).

Above all, as a mediator, the secretary-general is in no position to make credible threats or provide essential side-payments to the disputants. He is, more or less, a civil servant appointed by the member countries, and as an official put it, somebody whose salary is paid for by the member governments (Interviews in Brussels, July 2001). This fact is also acknowledged by the former Secretary-General, Lord Carrington, who maintains that "[…] the Secretary-General has no executive power. He cannot take initiatives, or give instructions to anybody actually to do anything, unless it be to a member of his personal staff to produce a paper. He can only act as the servant of an institution which operates by consensus. Everyone has to agree" (1988, 380). Rühl maintains, too, that no political power of initiative or decision-making is vested in the office of the secretary-general. He adds, however, that this does not preclude a political function for the secretary-general. The secretary-general, just like the SACEUR or the Chairman of the Military Committee, may exercise "an indirect power of

---

[63] Stikker maintains that he received a briefing on US foreign policy in different areas during his tenure (1965, 16). On the other hand, the French were quite uncooperative in the de Gaulle era and allowed him to read French political reports only once or twice in three years (Ibid., 18). This gives a hint at the difficulties facing the secretary-general.

Secretary-General Manlio Brosio, on the other hand, primarily relied on the newspapers as sources of information. In addition, he also received information from the permanent representatives (Jordan 1979, 183).

political advice and influence through personal persuasiveness, public statements, and diplomatic maneuvering between the Allied governments" (1974, 463). Moreover, the "power" of the office of the secretary-general seems to be dependent on the personality furnishing this office and his conception of the role to be played by the secretary-general. For example, while Lord Ismay is said to have acted as a manager of NATO's administration, his successor Paul-Henri Spaak viewed himself, according to Smith, as "Mr. NATO" and asserted certain authority for the office (Smith 1995, 67; Rühl 1974, 464). Smith claims that Spaak's Cyprus initiative can be viewed as an attempt by the Secretary-General to enhance the institution and consolidate its new role as a mediator in disputes among member-states (1995, 67). Manlio Brosio as well as Joseph Luns also wanted to be more than mere managers and tried to influence and bring the interests of the Alliance to bear on allied decisions (Rühl 1974, 466). Even so, they faced the same constraints that their predecessors did. The secretary-general, as Rühl puts it, can inspire action but is in no position to enforce such action. Rühl concludes that "[t]he most that can be made of this post is a spring-board for orientating the national powers in the desired direction, for mediating between opposing interests if the mediation is requested and its result accepted, for warning and educating governments if they are willing to listen. There is no real power in this office and there is no other power in NATO unless the major allies lend it- and they lend it on their terms" (1974, 466).

Overall, while the organization has entrusted the secretary-general with the right to bring inter-alliance disputes to its attention and offer his services to the disputants, he has neither a mandate, nor any means at his disposal to play the game of *carrots-and-sticks*. He can contribute to the resolution of conflicts by facilitating communication between the parties and putting forward proposals if the disputants allow him to do so. However, the tough job of moving countries to compromise by making use of threats and rewards is done by the bigger partners.

4.2) NATO's Record of Mediation in Greek-Turkish Disputes

NATO was interested in Cyprus for various reasons. First and foremost, the conflict on the island burdened relations among three member-states and had the potential to lead to a war between Greece and Turkey. Under such circumstances, the Soviet Union might intervene either directly or gain a foothold on Cyprus indirectly given the large communist minority there (Windsor 1964, 4). In the same manner, any of the parties might be inclined to pursue closer ties with Moscow in order to further its position vis-à-vis the other disputants. Moreover, the island was of strategic importance hosting US intelligence and communication facilities as well as British airbases. As Windsor notes, "NATO had therefore powerful reasons to offer what services it could in order to avoid conflict between Greece and Turkey and appease the situation in Cyprus" (Windsor 1964, 4).

In a first attempt to defuse tensions between Greece and Turkey, the commander of Allied Forces, Southern Europe (CINCSOUTH), Admiral William M. Fechtler was sent into the region in the aftermath of the 1955 anti-Greek riots in Turkey. His efforts complemented the activities of US Secretary of State Dulles who sent letters to the governments in Athens and Ankara and reminded them that they had to subordinate their conflict to the broader interests of the Alliance (cf. Stearns 1992, 33-34). Greece, nonetheless, responded by withdrawing its forces from an announced NATO maneuver, which set a precedence for the Greek strategy of downgrading its association with the Alliance whenever a crisis erupted with Turkey.

In 1956, Secretary-General Lord Ismay proposed that NATO could mediate between Greece and Turkey (Kay 1998, 52; Jordan 1979, 37). However, deliberations in the NAC failed to bring about a consensus that would have made such a mediation possible. Great Britain, Turkey, and Greece opposed the plan, the US had no inclination to pressure the British, and the French were busy focusing their attention on their North African endeavors (Jordan 1979, 37; see also Spaak 1969, 361). In the end, Lord Ismay had no choice and gave up his plans for the Alliance's intervention.

The successor of Lord Ismay, Secretary-General Paul-Henri Spaak maintains that his attention had been drawn to the conflict between Greece and Turkey when he took office, but he had not wanted to take any sensational steps, because a failure would have damaged the institution as such (Spaak 1969, 361). However, encouraged by Makarios to mediate in the dispute, Spaak sent a letter to the Turkish Prime Minister, Adnan Menderes, wherein he offered his mediation efforts to the parties on the condition the prospect for arriving at an agreement was present. Unless the parties showed some flexibility and stopped insisting on irreconcilable demands like enosis and taksim, an intervention by the Alliance made no sense and would help to undermine its authority, simultaneously underscoring the existing division lines within NATO (Spaak 1969, 362-363). Spaak proposed to start discussions on the basis of a conditional independence for Cyprus. The formula for independence should be drafted in a manner securing the rights of the minority and reflecting the interests of the Alliance. Furthermore, there should be provisions foreclosing the door to future demands for enosis or taksim.[64] Spaak explored whether the Turkish side would be ready to enter into negotiations under such terms.

---

[64] According to Woodhouse, what Spaak suggested was independence for Cyprus within the framework of the Commonwealth (Woodhouse 1982, 71).

As Ankara did not respond to Spaak's letters, Spaak thought his offer was neither accepted nor rejected and decided to continue his efforts (Spaak 1969, 364). He thereupon organized a confidential meeting with the representatives of Greece, Turkey, and Great Britain to discuss the issue in his residence where, in his view, essential progress was made (Spaak 1969, 365). However, they were to be overtaken by the announcement of a new plan drafted by the British Premier Macmillan. The plan called for two separate legislative councils for Greek and Turkish Cypriots and requested Greece and Turkey to appoint representatives subordinate to the British governor (for the plan see Ierodiakonou 1971, 172). The provisions of the plan were unacceptable to Greece. After visiting Ankara and Athens, Macmillan told Spaak that he was willing to make changes in his plan. However, he rejected the adoption of a proposal suggested by Spaak at an earlier stage – and deemed worth discussing by Karamanlis (Woodhouse 1969, 77) - according to which the presidents of the elected local councils should be the aides of the governor instead of representatives appointed by Greece and Turkey (cf. Spaak 1969, 366). On the other hand, what Macmillan offered was to curb the powers of the aides. When the planned amendments were made public, they again caused uproar in Greece. The Greek representative delivered a memorandum to the Secretary General, wherein Greece warned about the negative consequences the implementation of the Macmillan Plan - scheduled to come into effect on October 1st, 1958 - could have on the Alliance and on peace in the Eastern Mediterranean (cf. Spaak 1969, 367). On September 22nd, 1958, Karamanlis sent a letter to the Secretary General declaring that the imposition of a plan unacceptable to Greece would jeopardize the country's continued membership in the Alliance (cf. Spaak 1967, 367). Spaak thereupon summoned the representatives of all countries except for those from Greece and Turkey and urged them to support him in his attempts to move the British government to postpone the implementation of the plan. As Spaak did not get unequivocal support, he decided to go to Athens where he met Karamanlis and the Greek Foreign Minister Averoff. He was told in Athens that the Greek government had only two options if the plan was not to be abandoned or at the least postponed. They would either have to resign and cede power to the extremists or leave NATO as demanded by the public (Spaak 1967, 368) and Makarios (Woodhouse 1982, 79).

When Spaak returned to Paris, he proposed a new conference where the parties should try to find a temporary solution based on a modified version of the Macmillan Plan that he himself had drafted (see Spaak 1969, 368-369; Ierodiakonou 1971, 198-199). What followed were what Spaak called the most stormy discussions in the NATO Council since the institution's founding (1967, 369). In the initial stage, the Turkish representative accused Spaak of not being impartial and declared he no longer enjoyed the trust of his government. The representative added that Turkey would make its stance about the idea of a conference known after Britain had made its stance public. The Greek representative, on the other hand, maintained that his country could take part in such a conference.

The Council gathered again a couple of days later to continue discussions on the issue. The British representative declared his government could not accept any delay in the implementation of its plan (Spaak 1967, 370-371). However, Britain was ready to participate in a conference where the discussions would proceed on the basis of the Macmillan Plan. The following discussions in the NAC concentrated on the role to be played by the representatives of Greece and Turkey in the affairs of Cyprus and on the question of the basics and the scope of a conference if the parties agreed to this. The Council, in the end, succeeded in drafting a protocol to be submitted to the member governments (Spaak 1969, 372).

Greece proposed two changes in order give its consent to a conference. First, apart from working out an interim solution based on the plans put forward by the Secretary-General and Macmillan, the conference should also have the mandate to formulate a lasting and final solution as well (Spaak 1967, 373-374). Secondly, in addition to Greece, Turkey, Britain, and the representatives of the two communities on the island, the Secretary-General himself, and "impartial" parties such as the US, Italy, and France should also participate in the conference. Great Britain and Turkey, however, refused to accept the second proposal, whereupon Greece broke off the negotiations (Spaak 1967, 375).[65]

Spaak claims that while his efforts and the discussions in the Council did not bring about a final settlement, they were still useful as they helped to bring Greece closer to NATO, calmed tensions, and, above all, contributed to the signing of the London and Zurich accords of 1959/60 (1967, 359, 379).[66] This view is also supported by Stearns, who asserts that the efforts within the framework of NATO contributed to the resumption of talks between Greece, Turkey, and Britain leading to the conclusion of the aforementioned treaties (cf. Stearns 1992, 78; see also Papacosma 2001, 201).

---

[65] The fact that the Cyprus problem was still on the agenda of the UN and the discussions there might have brought forward results favorable to Greece (Spaak 1969, 378) as well as the pressure exerted by the opposition in Greece (Woodhouse 1982, 81) played a role when Karamanlis decided to suspend negotiations.

[66] After the accords had been signed, Spaak received telegrams from Karamanlis, Selwyn Llyod, and Christian Herter (Spaak 1967, 379). All stated that the agreements would not have been possible without his contributions and thanked him for his commitment.

In 1963, the system established by the treaties of 1959/60 broke down on Cyprus and inter-communal warfare followed. About 40,000 Turkish Cypriots had to leave their villages and resettle in enclaves to the north of Nicosia (Stikker 1966, 407). Secretary-General Dirk Stikker maintains that he felt a necessity to intervene since the increasing tensions could have entailed a war between the NATO partners Greece and Turkey, which in turn might have led to an intervention by Bulgaria and the Soviet Union, unilaterally or in concert, or, at least, caused a shift in loyalties in their quest for the support of third parties (Stikker 1966, 403-404). Stikker thereupon informed both parties that he was interested in assisting them in solving the conflict (Ibid., 407). He urged both parties to exercise restraint and appealed to Turkey not to intervene militarily on the island (cf. Jordan 1979, 136) The Turkish PM İnönü was receptive towards the requests made by Stikker and both countries stopped "provocative" military movements (Jordan 1979, 136; Stikker 1966, 408). Stikker asserts that he himself probably twice and General Lemnitzer at least once helped to prevent open hostilities during this period (Ibid., 408).

However, the crisis was still not ended. The Greek Cypriots continued to build up their military forces. Meanwhile, the UN Security Council sent peacekeeping troops to the island. In the face of the fact that the measures adopted by the UN failed short of ending the conflict and tensions continued to increase, Stikker decided to intervene once more. The Secretary-General visited Athens and Ankara, and went to the US to discuss the issue with the UN Secretary-General U Thant (Ibid., 408). Stikker's visit in Athens was complicated by the Greek officials who leaked the content of the negotiations to the press. Greece, in fact, wanted to prevent an intervention by NATO, once more preferring UN forums (Jordan 1979, 136) - nonetheless, it had been Greece which had asked the NATO Council on January 2nd, 1964, for support in preventing a military intervention by Turkey (Windsor 1964, 17).[67] The matter was taken up by the Ministerial Council of May 1964. The NAC reaffirmed its support for the UN mediator - meeting Greek demands - and approved the efforts of the Secretary-General and entrusted

---

[67] Before NATO could become active, it had to overcome internal disagreements over the role it was to play. Whereas the US, UK and Turkey wanted the Alliance to intervene, Greece and France were in general hostile to a NATO involvement (cf. Windsor 1964, 5-10). Other European members were not interested in getting involved in an extended colonial dispute either. Thus an intervention could have reinforced internal division lines and might have been viewed as an act in favor of Turkey. A policy of non-intervention, on the other hand, could have alienated Turkey, as it would have looked like a pro-Greek stance. Furthermore, Makarios did not want an intervention that could curb his control over the affairs of the island. Neither did the Cypriot Greek population endorse the idea of an intervention by NATO, which they viewed as an obstacle to self-determination. Moreover, Makarios had already called the UN and an intervention by NATO after a UN mediator had been appointed would have played into the hands of Soviet propaganda.

him with a watching-brief, calling upon him to keep a track of the developments and to consult the NAC whenever he deemed it necessary, which was a position favorable to Turkey (Jordan 1979, 136; Stikker 1965, 13; Stikker 1966, 408).

Meanwhile, Makarios visited Nasser and contacted Moscow. Stikker, on the other hand, decided to visit Ankara and Athens again to get the parties back to the negotiation table. When the Greeks - in an effort to avert an intervention by the Alliance (Jordan 1979, 136)[68] - informed him that PM Papandreou would not be able to meet him, Stikker, as he puts it, had to rely on "unorthodox"[69] means to

---

[68] According to Jordan, Secretary-General Stikker claimed he was no mediator but acted as a "watch-dog" in an attempt to defuse tension between the two NATO partners (Jordan 1979, 136).

[69] Elsewhere Stikker further elaborates how he had managed to arrange an appointment with Papandreou (Harry S. Truman Library: Oral History Interview with Dirk U. Stikker, http://www.trumanlibrary.org/oralhist/stikker2.htm#34, downloaded on May 3rd, 2004, 9 p.m.):

> STIKKER: Well, I think that the United States administration was rather divided. As always, George Ball was the strongest man making propaganda for supranationality. Years later, when I was in NATO, and when we had difficulties between Greece and Turkey and war was imminent, at that time I went to Greece and when I was there I had long discussions with [George] Papandreou. He had to do something, and I couldn't accept it. I had great difficulties in reaching Papandreou. He didn't want to receive me, and I used again personal relations to make him take part in the discussion. And I used Onassis. I said.to Onassis, "If I'm well informed, you've just transferred all your money to Greece at the present moment. You haven't got it any longer in the Argentine, or wherever your money is, but you have quite an important interest now, money wise. What do you think will happen to your money when war occurs between

> [35]

> Turkey and Greece?"

> And he said, "Is it serious as that?"

> I said, "Yes, it is."

> He said, "I'll ring you up in a quarter of an hour."

> He got back to me. We were in a plane together when I told him this story. One of the Olympic athletes, on of his friends, said, "What is happening to this plane?"

> And he said, "When we land I will ring you up in a quarter of an hour. If Papandreou doesn't want to receive him, I've got to interfere."

> In a quarter of an hour in my hotel, he rang me up. He said, "Papandreou asks you to dine with him tonight. I'll be there also."

arrange an appointment with the Greek PM (Stikker 1966, 409). According to Stikker, the Greeks now recognized that something had to be done and the option of the "NATOification" of Cyprus through enosis was one of the issues discussed (Ibid., 409). Not providing any further information on the content of the talks, Stikker reports to have consulted the Americans afterwards whose "response" was encouraging. Stikker was joined by the US Undersecretary of State George Ball the next day. After discussions between Ball and Greek officials, both went to Ankara where they continued their mediation efforts. However, they were not able to put an end to the conflict. After this last mission also "failed," Stikker declared there was nothing more he could do and "was content to stay abreast of the affair from his Paris office" (Jordan 1979, 138). He did not want to impose himself on the parties and alienate either Greece or Turkey. In retrospect, Stikker concludes that despite the fact that neither the Greeks nor the Turks immediately accepted the proposals put forward by him, two important things had been achieved (Ibid., 410). The Greeks had been persuaded that they had to act if they wanted to prevent a disaster, and the Americans were now willing to play a more prominent role, which became visible with the arrival of George Ball and was reaffirmed by Dean Acheson's participation in the Geneva talks later on.

Regarding the developments of 1964, the question arises why Turkey refrained from intervening on the island despite threatening to do so many times. Some authors attribute Turkish *inactiveness* to the harsh warning by President Johnson who stated in a letter to the Turkish Prime Minister, İnönü, Turkey could not count on the NATO security-guarantee in case of Soviet aggression. It was this threat of abandonment in the view of Ronald Krebs, for example, that prevented Turkey from intervening (1999, 361). However, there seem to have been a number of factors making a Turkish intervention almost impossible in 1964 regardless of the US threat.

In the first line, in the view of certain analysts, Prime Minister İnönü - described by Riemer as a cautious character in person (Riemer 2000, 70) - did not really want a military invasion in Cyprus (cf. Uslu 2000, 94, 107-108). İnönü had doubts whether the Turkish military was able to carry out such an operation with success. Moreover, the Soviet Union was publicly supporting Makarios and still wanted to see the independence of the island preserved, opposing any intervention by NATO. İnönü feared that the Soviets could side with Makarios in the event of war.[70] In addition, the Turkish Prime Minister was interested in maintaining close ties with the US and was concerned about the prospect of a possible US arms embargo. As a consequence, İnönü is said to have informed the US before giving the green light to the operation in anticipation of a negative US reply. He hoped that the US would intervene and prevent Turkey from carrying out its plans. In such a case, the US would serve as a scapegoat and could be made responsible for not intervening. Thus the anger of the public and the hawks in the military could be redirected towards the US. In this connection, Uslu points to statements by former officials close to İnönü confirming this view (cf. Uslu 2000, 109). For example, he cites the case of a former Turkish diplomat stating that the Turkish army had not been in a position to carry out such an operation and that PM İnönü had been well aware of this fact. However, he would have the military prepare for invasion and sail out towards Cyprus in order to send a warning to the Greeks. In the end, the US would step in, which would enable the Turkish government to redirect criticism towards Washington.

Overall, the unwillingness of the Turkish Prime Minister to order a military operation, the fact that the Turkish forces were not in a position to carry out such a mission yet, and the opposition aired by the Soviet Union, seem to have been the main reasons behind the Turkish decision not to invade the island. In fact, since Turkey had no real desire to intervene militarily, a clear message short of such a warning of abandonment pronounced in the "Johnson Letter" would probably have been sufficient as well. It was not the US opposition which had surprised the Turks, but the harsh wording of the letter (see Uslu 2000).

Leaving aside the question of what hindered Turkish military intervention, what is surprising about the developments of 1963/64 is that NATO would have almost gone out of area for the first time in its history. To be more precise, when negotiations did not bring about any solution and the communal strife continued, Britain suggested the creation of a NATO peacekeeping force to be deployed to the island in order to avert a Turkish intervention (Nicolet 2001, 316-317). Nonetheless, the British soon realized that the prospects for the implementation of such a project were dim. First, many NATO partners were likely to oppose such a force, rendering its creation in the short term, if at all, impossible. In addition, the NATO forces had not been trained for such contingencies. Furthermore, since

---

[70] The Soviet stance adopted in the aftermath of the Turkish air raids flown against Greek Cypriots in August 1964 proved that such concerns were warranted. In response to Makarios' requests for support, the Soviet Union had declared that it would support the Cypriots if any external powers tried to invade the island and maintained that talks on the issue could begin immediately (cf. Ierodiakonou 1971, 260-261).

Cyprus was not a member-state, the deployment of NATO forces required the prior invitation of Makarios, which seemed impossible anyway (Nicolet 2001, 3116).[71]

As an alternative, the British suggested the creation of a peacekeeping force of about 10,000 men to be drawn from NATO member-countries, though not under the command of NATO (Nicolet 2001, 317; Windsor 1964, 12; Stearns 1992, 35, 78; Uslu 2000, 64). The US, first negatively tuned towards the idea, later offered its support to the plan. The issue was finally discussed in the NAC. The French rejected the plan outright while the Germans were to announce their opposition after Makarios had rejected the plan (Windsor 1964, 13). Makarios wanted to avoid any measures that could jeopardize the non-aligned status of the island and feared that the Greek Cypriot interests might be sacrificed for the interests of Turkey and NATO (Stearns 1992, 78; Papacosma 2001, 203). Particularly the fact that the force was to encompass 10,000 men, quite large for a tiny island like Cyprus, and that the Cypriots themselves would not be included in the international committee to be established to give political guidance to the force added to suspicions in Cyprus (Joseph 1995, 237). There were fears that the aim of the force went beyond restoring peace and order, and the force came to be viewed almost as an occupation army (Joseph 1995, 238). Given this opposition of Makarios, the idea of creating an international peacekeeping force within the framework of NATO ultimately had to be abandoned. Instead, a UN force was installed on the island that failed to guarantee the safety of the Turkish Cypriots. Thus the necessity for NATO to continue its search for ways to defuse tensions remained intact. This time, however, the efforts were limited to furnishing the secretary-general with a watching-brief.

The NAC continued to be concerned with the developments on the southern flank and approved the watching-brief in its first ministerial meeting after Manlio Brosio assumed office as NATO's new Secretary-General in 1964 (cf. Jordan 1979, 210). Only two weeks after taking office, the fly-over of Cyprus by Turkish

---

[71] An alternative strategy would have been to bring Cyprus into NATO before deploying forces (Nicolet 2001, 317). Indeed, the issue of Cypriot membership had already been discussed when Cyprus became independent. According to Nicolet, Greece and Turkey had signed a secret gentlemen's agreement during the London Conference concurring to work towards Cyprus' admission to NATO in due time (Nicolet 2001, 314). The issue was also discussed among US and British officials when Makarios asked the US for military aid for the army to be established. Nonetheless, the US and the UK concurred that a Cypriot membership would entail more disadvantages than benefits and thus opposed such a move (Nicolet 2001, 314-317). The issue of providing military aid to Cyprus did not require any further consideration when the idea of a Cypriot army was abandoned.

fighter jets induced Brosio to send a letter to the Turkish PM İnönü inquiring into the issue and offering his offices to contribute to the settlement of any disputes. In 1965, Brosio was to play an important role in bringing together the foreign ministers of Greece and Turkey during a Council meeting for the first time in sixteen months (cf. Ibid., 410).

A quite tranquil three-year period in relations between Turkey and Greece was to end in 1967 when inter-communal fighting flared up, and Turkey started to prepare for an intervention again. While the Americans sent Cyrus Vance into the region to mediate, Secretary-General Brosio conferred with the permanent representatives and appealed to both countries to show restraint (cf. Jordan 1979, 217). When tensions escalated, Brosio suggested he might visit both capitals and try to defuse tensions (see also Kay 1998, 53). The Council accepted his proposal, and Brosio left for the region together with his political adviser. There, he met Cyrus Vance who was at first hesitant to work in concert with the Secretary-General given the limits of his office (Jordan 1979, 211). However, after the meeting, they decided to work together and coordinate their actions. According to Jordan, Brosio did not play an independent role, and Vance could have fulfilled the mission alone, too (Jordan 1979, 212). In a similar manner, Kay maintains that despite the fact that both had worked in a complementary manner, it was again the US which had prevented war (1998, 53).[72] Brosio's participation, however, "made the settlement somewhat more palatable to the disputants" (Jordan 1979, 212).

When Secretary-General Joseph Luns was informed that a coup had been carried out against Makarios in 1974, he summoned members of the Greek and Turkish delegations to his office and urged both parties to exercise restraint (cf. Tülümen 1998, 136). On July 15th, 1974, the Secretary-General sent letters to Greece and Turkey reiterating his appeal to both sides to show restraint (NYT, July 17th, 1974). The NAC gathered many times to deliberate on the issue on the following days. It announced support for Makarios and asked both parties to show restraint, while urging Greece to withdraw its troops from Cyprus (see also Varvaroussis 1979, 99; NYT, July 18th, 1974). Greece, on the other hand, kept denying any involvement in the coup and stated it was supporting the territorial integrity and independence of Cyprus. During another NAC meeting, Greece declared it was

---

[72] This time US pressure would come to bear especially on Greece. As stated by Uslu (2000, 215), it was now the turn of the Greeks. Cyrus Vance informed them that the US would not step in to prevent a Turkish invasion this time (Ibid., 216). As the junta had no good standing anyway, given its authoritarian character, it could not dare to further worsen its position by waging a war with a NATO partner. They accepted the need to withdraw their forces from Cyprus and called back Grivas (see also Sönmezoğlu 1995, 26).

ready to replace its 650 officers in the Cypriot National Guard (NYT, July 19th, 1974). Such a move was of course short of satisfying Turkish expectations, and Turkish troops landed on the island on July 20th, 1974.

Following the invasion of Turkish troops, the Council met for an emergency meeting to discuss the developments (Tülümen 1998, 142-145). These discussions were to continue in the following days. The Greek representative asked the allies to help put an end to the invasion. His Turkish counterpart, on the other hand, accused the Greeks of providing false information to the Council and made public the conditions defined by his government to prevent further escalation (Ibid., 145). Turkey was willing to accept a ceasefire on condition the safety of the Turkish troops would be guaranteed. Thus, Secretary-General Luns asked the Council to focus its attention on the issue of reaching an agreement on a ceasefire. Thereupon, the Greek representative declared his country did not want a war with Turkey and said he would consult with his government on the issue. Thereafter, Luns sent a message to the governments of both countries stating that the Council had almost unanimously embraced the idea of a ceasefire and urged both countries to accept a ceasefire as suggested by the US president for July 22nd, 2 p.m., GMT (Tülümen 1998, 147; see also NYT, July 23rd, 1974). Both countries responded positively to the Secretary-General's appeal and the Council was informed about the developments during another meeting on July 22nd, 1974 (NYT, July 23rd, 1974).

On July 23rd, 1974, a crisis erupted between the UN, Britain, and Turkey (Tülümen 1998, 149). The British had formed a peacekeeping force and taken over the airport of Nicosia from the Greek Cypriots. Now Luns was asked by the British Foreign Minister Callaghan to pressure the Turks not to try to take over the airport. In the meantime, the Secretary-General of the UN, Kurt Waldheim, told the Turks the UN forces were authorized to defend themselves if the Turks started an attack (Ibid., 150). Thereupon, Secretary-General Luns sent a message to the Turkish delegation urging the Turkish government to refrain from launching an attack against the airport. The crisis was defused when Turkey gave assurances that its forces would not try to take over the airport (Birand 1976, 152).

When the Geneva talks were adjourned on July 30th, 1974, Karamanlis requested Secretary-General Luns[73] to summon a meeting of the NATO foreign ministers. The Secretary-General, however, replied that this was not possible since he and

---

[73] When an arms embargo was imposed on Turkey, Secretary-General Luns was to travel to Washington to persuade the members of Congress to be more considerate towards Turkey and continue to provide military aid to Turkey (Rühl 1976, 24).

most of the foreign ministers would not be able to attend such a meeting because they were either too busy or on holiday (cf. Woodhouse 1982, 216; see also Stearns 1992, 68; Papacosma 2001, 204).

The second round of the talks in Geneva ended in a deadlock on August 13th, 1974, and some hours later Turkey started the second phase of the invasion.[74] Thereupon, Karamanlis, having been told by his chiefs of staff that no Greek military response was possible, ordered Greece's withdrawal from NATO's integrated military structure only four hours after the Turkish invasion had begun (cf. Woodhouse 1982, 217-218). Greek officials maintained that this move was "aiming at arousing world opinion and bringing it to bear against Turkey" (NYT, August 15th, 1974). The Council met in an emergency meeting and Secretary-General Luns broke off his holiday in the Black Forest to attend the meeting (NYT, August 15th, 1974). Luns now proposed to visit Ankara and Athens. However, his offer was rejected by both parties.

Greek foreign minister Mavros declared NATO had lost its *reason d'être* given its failure to prevent conflict between two members (Varvaroussis 1979, 112). The Greek policy-makers thought that Greece's withdrawal would force the United States to abandon its pro-Turkish stance and isolate Turkey cutting its strategic and military links to Western Europe.[75] As a price for Greece's return, the allies might be willing to take some initiative on Cyprus, Karamanlis hoped (Woodhouse 1982, 219). Another reason behind the decision to leave the command structure was the expectation that such a move would help to satisfy the expectations of the public, and especially of the armed forces – the decision was indeed received with euphoria in the population. The Greek Prime Minister declared later that he had to choose between two options. He either had to declare war on Turkey, or leave NATO. He chose the "lesser evil" (cf. Stearns 1992, 68; see also Papacosma 2001, 203). On the whole, the withdrawal helped to avoid an "explosion," as Varvaroussis calls it (1979, 114; see also Meinardus 1982a, 208), which could have led to a war against Turkey.

---

[74] Birand claims that the junta had decided to transfer the rule to a civilian government prior to the demands made by the chiefs of staff to back down. They had been anticipating that the civilians would have to make compromises vis-à-vis Turkey, which would in turn enable them to blame the civilians for having sold Cyprus to the Turks and recapture control again (1976, 146). Thus Karamanlis was in no position to accept any proposals put forward by Turkey. Turkey, on the other hand, wanted to solve the issue once and forever and was not willing to accept anything less than a cantonal or bi-zonal federation. According to Birand, it was mainly for these reasons that the conference had been stillborn (Ibid., 216).

[75] Indeed, the communication lines connecting Turkey to Europe went through Greece and thus Turkey was cut off from NATO when tensions arose with Greece (Jordan 1979, 178)

However, while the Greek withdrawal helped to appease public sentiments, it did not have the desired effect of isolating Turkey. To the contrary - except for the fact that Greece's decision to leave NATO's military arm earned Greece the favor of non-aligned countries, a move that would benefit Greece in the UN - it had the opposite effect of adding to the strategic and diplomatic weight of Turkey (Meinardus 1982a, 209, 405). The Greek government was to realize this fact within a couple of months after the decision to leave had been taken and started to pursue a reintegration of its forces into NATO giving up a prior claim that Greece would only return after the Turkish troops had left the island. Coupled with this desire to return to NATO, in the post-1974 period, Greece intensified its strategy of internationalizing the conflict and indirectly pressuring Turkey (Meinardus 1982a, 213). In the end, Turkey agreed to Greek reintegration in October 1980.

In retrospect, both countries seem to have been closer to war in 1974 than at any time before. The leaders of the junta in Greece had in fact decided to go to war against Turkey, but could not do so because of the resistance of the generals and because the mobilization of the forces failed given the catastrophic state of the armed forces. According to Bolükbaşı, the Turks, on the other hand, had been expecting that Greece would declare war (Uslu 2000, 279). Thus 70 percent of the Turkish troops and material had been dispatched to the area bordering the Aegean while only 30 percent were being used in Cyprus, which rendered the *mission* there more difficult. Nevertheless, this time the Americans refrained from sending a letter with the wording of the one sent in 1964. Convinced that they could not stop Turkey, the Americans chose to refrain from measures that could have further alienated Turkey. There was no willingness to put relations with Turkey at risk once more. Kissinger was convinced that the US could not prevent Turkey from carrying out its plans this time, and thus threatening Turkey while there was no prospect for success made no sense and could only lead to an unnecessary deterioration in their bilateral relations (Uslu 2000, 275-277).[76] The *New York Times* (August 16th, 1974) reported that Kissinger had been telling visitors in Washington he could not pressure Turkey beyond a certain point, since "US strategic interests in Turkey were too important to jeopardize." Another reason why the US might have refrained from pressuring Turkey so much might have been the expectation that it would not be so hard to avert a Greek-Turkish war, since the Greek military was not prepared for a war against Turkey (Uslu 2000,

---

[76] The US was in no position to stop the Turkish intervention because it had to deal with the Watergate scandal (Uslu 2000, 297). By the time the second Turkish invasion was carried out, President Nixon had already resigned and Kissinger was busy forming a new administration. James Callaghan, the British Foreign Minister at the time, maintains that Kissinger was so busy with the Watergate affair that he lost contact with him during "critical hours" (Callaghan 1987, 339).

277). Nonetheless, the Americans threatened to cut the assistance to Turkey, but the aid issue could not be used as an effective measure this time, because the aid levels to Turkey had been reduced anyway when the Turkish government had lifted the ban on the plantation of poppy (Uslu 2000, 273). It was only after the invasion had begun that Kissinger threatened Turkey that the US would remove all nuclear weapons if it did not agree to a cease-fire (Callaghan 1987, 345). Whatever role this threat might have played in the calculations of the Turks, the fact remains that they agreed to a cease-fire on July 22nd, 1974. It was only in February 1975 that the Congress imposed an arms embargo on Turkey despite opposition by the Administration against such a move.[77]

Both countries were to find themselves at the brink of war against in 1987 when differences related to the delineation of the continental shelf led to a new crisis. The NAC discussed the developments during a meeting on March 27th, 1987, and urged both parties to solve their differences peacefully (cf. Axt & Kramer 1990, 95). Two days later, NATO Secretary-General Lord Carrington offered his mediation efforts to both parties (Ibid., 43). While the offer was accepted by Turkey, Greece rejected it, stating it preferred other channels like the ICJ. Indeed, when the crisis had erupted, Greece had sent its foreign minister to Bulgaria to ask for their support. Moreover, the Greeks had chosen to inform the countries of the Warsaw Pact and Islamic states before the ambassadors of the NATO partners

---

[77] Yet, the decision to impose an embargo had neither been the result of an objective consideration of the situation. The Greek lobby in the US and its representatives in the Congress had played an important role in the adoption of the embargo strategy (see Varvaroussis 1979, 187). Apart from the influential role played by the Greek lobby, the decision to impose an embargo in spite of the opposition voiced by the administration against such a move had been the result of a desire by the Congress to reassert its position against the administration after the debacle in Vietnam (Meinardus 1982a, 220). Uslu, too, states that the embargo issue served as a welcome opportunity for the Congress to reassert its role in the foreign relations of the country (Uslu 2000, 329; Sönmezoğlu 1995, 90).

The embargo issue was to be used by the US as leverage to pressure the Turks to make concessions on Cyprus. However, the US began to change its strategy by March 1978 and US officials hinted that the embargo might be lifted (Tülümen 1998, 230). Tülümen contributes this change of mind and the eventual lifting of the embargo in September 1978 to the developments in Iran (cf. 1998, 227-231). After Turkey had shut down US installations in Turkey, two intelligence installations had been set up in Iran to serve as substitutes. Iran had been central to the US strategies in the Middle East anyway and served as a bulwark against Soviet expansionism in the region. Nevertheless, the strengthening of Ayatollah Khomeini's opposition against the Shah's regime and the incidents of January 7th, 1978, leaving a number of students dead and marking the beginning of the Islamic Revolution in Iran convinced the US that Iran might soon be lost, and this forced it to redirect its attention towards Turkey. Despite the fact that the Cyprus issue had not been solved yet, the US had now chosen to lift the embargo. It was necessary to mend ties with Turkey before Khomeini seized power in Iran.

were informed by the deputy foreign minister (Axt & Kramer 1990, 42-43). In the end, it was pressure exerted by NATO partners US, UK, France, and Germany that persuaded Turkey to declare that its vessel would remain within the boundaries of the Turkish territorial waters.

When a dispute over the status of some islets in the Aegean brought the NATO partners to the brink of war in 1997, NATO and the US intervened once more. In a meeting of the NAC, both countries were urged to put an end to the dispute (Kay 1998, 53). Secretary-General Javier Solana's offers for good-offices were rejected by Greece which claimed that the status of the islets was not negotiable. It was again intervention by the US President, Bill Clinton, and his aides that put an end to the crisis and prevented an intra-NATO war (see Kay 1998, 53).

During the S-300 crisis, Greeks and Greek Cypriots stated that they could refrain from deploying the missiles if a no-fly zone was established over Cyprus and enforced by NATO or the US. In response, Secretary-General Solana was reported to have declared that NATO was ready to monitor such a no-fly zone (cf. TDN, July 20th, 1998). The Turkish side, however, rejected the idea, claiming that the missiles could not be made the subject of negotiations. Turkey's representative at NATO, Onur Öymen, declared NATO involvement on the island was out of question (TDN, July 29th, 1998). Washington did not welcome the idea of a compulsory zone either and suggested a voluntary moratorium on military flights. Pentagon spokesman Kenneth Bacon stated that it was very difficult for any NATO member to get involved in enforcing a no-fly zone that involved other NATO countries (TDN, July 20th, 1998).

There had also been rumors that "NATO might be predisposed to "force" a solution in Cyprus by admitting a united Cyprus as the 17th member of the Alliance" (TDN, September 20th, 1997). Secretary-General Solana, however, put an end to these rumors and made clear that Cyprus would not become a member in NATO. He added he supported the implementation of CBMs and was discussing such CBMs with Greece and Turkey "just about every day."

4.3) Confidence Building Measures (CBMs)

Secretary-General Javier Solana undertook efforts to contribute to an improvement of the climate in the Aegean as soon as he assumed office in 1996. While his offer of good-offices during the Imia/Kardak crisis was rejected by Greece, he continued pressing for the implementation of confidence building measures (CBMs) between both countries and reiterated he was ready to mediate

(TDN, May 22nd, 1996).[78] During his visit to Athens in March 1996, Solana stated, "NATO should operate as a peace mechanism to solve differences" and added that he would do everything he could in that direction (ANA, March 2nd, 1996). The Americans supported the Secretary-General's efforts and sent a high-ranking delegation to Ankara and Athens in an attempt to defuse tensions. Washington stated it wanted both countries to implement a number of CBMs in order to avoid renewed crises in the Aegean (ANA, March 29^{th,} July 3rd, July 5th, 1996).

After meeting with Turkish officials in May 1996, Solana declared he had "proposed certain ideas, which he hopes will have a beneficial effect on Greek-Turkish relations" (ANA, May 24th, 1996). According to Turkish sources, Solana had proposed monitoring military exercises by AWACs, the establishment of a hotline between Ankara and Athens based in NATO headquarters in Brussels, and the expansion of the 1998 protocol on CBMs. By July 1996, an agreement had been reached between both countries to implement one of the measures included in the 1988 memorandum, which both had signed but never implemented. In accordance with this agreement, both countries were to suspend all military exercises in the Aegean in the period July-August (ANA, July 5th, 1996). A year later, in July 1997, Turkey declared it accepted and welcomed the CBMs proposed by the Secretary-General and would temporarily limit its military activities in the Aegean in a unilateral move; that is to say no maneuvers would be held between July 1st and August 15th (TDN, July 3rd, 1997). Athens welcomed the move and declared it was already implementing such measures. Meanwhile, Athens rejected a Turkish proposal to exchange information on flights over the Aegean, claiming that the right to control flights already belonged to Greece.

After four months of discussions between Greek and Turkish officials with the participation of Secretary-General Javier Solana, a decision was reached in 1998 to revive the Memorandum of Understanding and the Guidelines for the Prevention of Accidents and Incidents on the High Seas and International Airspaces signed in May and September 1988 as a whole (TDN, July 29th, 1998). The officials maintained that the talks within the framework of NATO would continue, "in order to clarify where necessary and to strengthen and complement

---

[78] Turkey had welcomed Solana's appointment as Secretary-General on the grounds "this would make the person holding this post more attuned to the problems of the region" (TDN, May 22nd, 1996).

where possible the confidence building measures which the 1999 agreements provided for" (NATO Press Release (98)74, June 4th, 1998).[79]

The successor of Solana, Secretary-General Lord Robertson, announced on December 6th, 2000, that the talks between the permanent representatives had given rise to a new agreement (NATO Update, December 6th, 2000; see NATO website). Greece and Turkey had agreed to inform each other about the exercises scheduled for the following year within the framework of NATO's annual exercise conference. In May 2003, Greek foreign minister Papandreou declared that Greece and Turkey might sign a new agreement on CBMs envisaging two chapters, while the first one would be implemented within the framework of NATO (TDN, May 27th, 2003). Two months later, both countries decided to adopt two further CBMs within the framework of NATO. The first of the CBMs envisaged cooperation between their Defense Colleges in the field of science and on issues such as military doctrines, crisis management, or peacekeeping (NATO Update, July 23rd, 2003). Furthermore, they agreed to send personnel to the PfP Training Center of the other party for training purposes. The statement said the parties would continue talks in order to be able to achieve agreement on further CBMs on the basis of the list agreed upon earlier.

The value of such CBMs is, of course, not uncontested. When it was announced that an agreement would be signed on the implementation of all CBMs decided upon 10 years earlier, the *Turkish Daily News* reported that some observers in Turkey were playing down the importance of the agreement (cf. TDN, June 6th, 1998). They regarded it simply as a repetition of pledges made earlier but never kept. Axt and Kramer draw attention to the ambivalent character of CBMs as well. On the one hand, they might contribute to an improvement in the atmosphere enabling both countries to touch on thorny issues. However, on the other hand, they might just serve as alibi measures, because the countries do not dare to tackle the contentious issues (1990, 116).

A NATO official maintained that the radar pictures (recognized air picture) of flights in the airspace over both countries were being transmitted to the NATO headquarters in Naples, a measure that had been adopted about 3 or 4 years previously (Interviews in Brussels, July 2001). Whenever allegations arose, it was possible to see who had been flying when and where. As a consequence, there had been a decline in the number of such allegations. The Turkish daily *Radikal* reported, too, that allegations of airspace violations had almost disappeared since the aforementioned measure was adopted (Radikal, June 13th, 2003). Another

---

[79] According to Kramer, these measures were regarded as necessary by NATO officials in order to be able to establish the new command structure in the region (Kramer 2000, 172).

report by *Radikal* said that Turkey had started to provide information to Greece on the number and routes of flights carried out by Turkish planes in December 2001 using NATO as an intermediary (Radikal, February 7th, 2002). According to the report, Turkish planes had even started responding to the questions of the Greek flight control towers. Nevertheless, despite these positive developments, the effectiveness of these measures proved to be dependent on the goodwill of both parties involved when renewed tensions related to the airspace arose in summer 2003. Greece complained about increasing airspace violations and claimed that a civilian plane had been harassed by Turkish military jets, yet Turkey rejected the allegations, and the Turkish press accused Greece of creating virtual tensions.

The fact that the countries have only managed to agree on the implementation of four CBMs within the framework of NATO since the agreement reached in 1998 to implement all measures agreed upon in ten years prior gives rise to further doubts as to whether the CBMs can effectively contribute to an improvement in relations. Nonetheless, they have, at least, a declaratory effect and reflect both countries' willingness to further improve relations. Another gesture of this goodwill was the decision of the NATO ambassadors of both countries to hold a joint-reception to celebrate the 50th anniversary of their accession to NATO.

5) NATO AND REGIME CONSEQUENCES

5.1) Institutional Capture and Issue Linkage

It seems almost logical that two countries in conflict can hardly isolate their interactions within an institution from the usual course of their relations. The Greek-Turkish case within the framework of NATO obviously constitutes no exception to this rule. On many occasions, Greek-Turkish disputes had their repercussions on the Alliance and both countries tried to employ the institution to further their positions vis-à-vis the opponent. Moreover, the Alliance provided for many opportunities to create linkages, though not in order to facilitate cooperation but to punish the opponent. As argued by Ronald Krebs (1999), at times the Alliance gave both countries more to fight about, which led to a deterioration of their relations. The task here will be to enlist a number of such cases where the disputants *captured* the institution and employed *destructive linkages*.

Greece began to use the Alliance as a means to display its protest against Turkish policies or to gather diplomatic support against Turkey at a very early stage. It was after the anti-Greek riots of 1955 that Greece withdrew its forces from the NATO headquarter in İzmir and cancelled its participation in maneuvers for the first time. This strategy of disrupting bonds with the Alliance as a result of problems arising with Turkey achieved its peak with the withdrawal from the military structure in 1974. One of the reasons conditioning this behavior had been the expectation that its allies would take Greece more seriously and pay more attention to its interests. Greek decision-makers had hoped the Western states, especially the US, would put more pressure on Turkey in order to prevent a further alienation of Greece from the Alliance or to allow for Greece to be reintegrated into the military structure. This strategy backfired first as the Greek withdrawal added to the strategic importance of Turkey, and the command and control rights over the Aegean were transferred to Turkey. Thus Greek decision-makers were quick to realize that membership in the alliance was indispensable as long as Turkey remained a member. Staying outside while Turkey was still a member could only weaken the position of Greece vis-à-vis its foe on the other side of the Aegean. Greece, they thought, fared much better vis-à-vis Turkey when it remained in the Atlantic Alliance. Thus the Greeks decided to seek reintegration (see Meinardus 1982a, 209; McDonald 1988, 76).

However, Turkey now profited from the fact that a reintegration into the command structure could not proceed without its consent. Thus, Turkey made such a consent contingent on a redistribution of command and control rights in the Aegean and refused to accept a return to the status quo ante 1974 (cf. Meinardus

1982a, 407). In response, Greece turned to the US and linked the future of the US installations in Greece to a re-admittance of Greek forces into the NATO structure on terms acceptable to Greece (Meinardus 1982a, 435). In 1980, the Greeks explicitly warned Washington that all US military installations would be shut down unless the reintegration issue was solved and drew a clear link between the completion of a new base agreement and the country's re-entry (Laipson 1991, 169; NYT, October 22nd, 1982). At the same time, the Karamanlis government pointed to the leftist opposition party PASOK, which was opposing Greece's return to NATO, at least in rhetoric (cf. Meinardus 1982a, 461). Karamanlis argued that given public sentiments hostile to NATO and the opposition exercised by PASOK, Greece was not in a position to make concessions. If the allies wished for a Greek re-entry, Turkey had to bow down before the next elections to be held in 1981.

The fact that the Turkish veto was jeopardizing US interests in the region in connection with the prospect of a leftist government in Greece opposing NATO membership must have convinced the US that Greek reintegration had to be completed before the 1981 elections in Greece. In Ankara, a military junta was now in charge of affairs. Interested in improving its image in Washington and Europe, the junta was more forthcoming towards wishes coming from Washington. In the end, both countries agreed on a plan put forward by General Rogers (see above), according to which the reintegration of the Greek forces was to take place prior to a settlement of the issue. This can indeed be seen as a success for Greek diplomacy, since the Turks had given up their prior claim that this could happen only after Greece had agreed to a redistribution of command and control rights in the Aegean.

The NATOification of bilateral conflicts with Turkey even constituted the centerpiece of Andreas Papandreou's strategy towards that country. During a meeting of NATO defense ministers in 1981, Papandreou, who had claimed in 1979 that Greece would leave NATO if he won the elections (Meinardus 1982b, 111), asked NATO to grant security guarantees to the country against the Turkish threat (see also Krebs 1999, 365). When the NATO partners rejected the proposal, Papandreou refused to sign the final communiqué. Thus, the Alliance was unable to present a communiqué at the end of a meeting for the first time in its history (cf. Meinardus 1982a, 474-475).[80] Papandreou continued to insist on border-

---

[80] After Papandreou declined to sign the communiqué, he was welcomed at home as a hero (Mackenzie 1983, 5). Mackenzie describes such moves by Papandreou as "propagandist." In fact, before leaving Brussels, Papandreou told reporters that the process of Greece distancing itself from NATO had begun. However, according to reports by Reuters, Papandreou had assured his colleagues that Greece would remain in NATO (Meinardus 1982b, 113). It must have been

guarantees against Turkey. However, the Alliance did not change its stance and declared it could not recognize a Turkish threat (Ibid., 477). In the end, Papandreou gave up his demand for NATO security guarantees. This did not mean that he abandoned his strategy of using NATO forums for his campaign vis-à-vis Turkey (Ibid., 481). When Turkish jets penetrated air space claimed by Greece in May 1982, Papandreou was to cancel Greece's participation in a NATO maneuver. Later that year, he handed out a document listing the Turkish violations of Greek airspace during a NATO ministerial meeting (Clogg 1991, 18). In addition, he made Greece's cooperation with the Alliance contingent on the latter's ability to put an end to "Turkish provocations" (Meinardus 1982a, 482). Should the Alliance fail to do so, Greece would abstain from cooperation. Meinardus concludes that this strategy of refusing cooperation due to conflicts arising with Turkey was central to Athens alliance policy in the period 1981-1983 (1982a, 483).

Indeed, by 1984, Greece had declared that the main threat to its security did not derive from the north but from the east, that is to say not from the Soviet bloc but from its NATO ally Turkey. Veremis highlights this point and maintains that Greek decisions related to the continued membership in NATO and the purchase of military material as well as the fate of the US installations in the country have to a large extent been influenced by the Turkish factor (Veremis 1998a, 148). As

---

obvious to Papandreou that the Alliance would have never been able to grant such a guarantee. Since the Alliance works by consensus, Turkey would have had to accept this as well, which, of course, was impossible. Thus Papandreou's move was probably motivated by a desire to draw the allies' attention to the conflict and simultaneously gain the favor of the Greek public.

Loulis describes Papandreou's foreign policy as a function of Greek domestic politics (cf. Loulis 1984/85, 380-382). Papandreou's foreign policy was conditioned by a desire to appease the PASOK party-activists, whom Louilis describes as the "watchdogs of orthodoxy;" by the necessity to appease the communists, which induced Papandreou to adopt pro-Soviet policies and anti-Western rhetoric at times in order to prevent the communists from using their influence in trade unions to challenge the economic austerity plans of the government (see also Veremis 1988b, 259); and finally, by populism, so that he could capitalize on anti-American and ultra-nationalist sentiments in the public to divert attention from domestic problems. With regard to Papandreou's strategy to use foreign policy as a means to redirect the attention of the public towards external matters, Loulis cites the example of the 1982 general strike of bank employees where Papandreou "dramatized disagreements with NATO" (1984/85, 382) and urged the employees to consider the fact that Greece was facing an international crisis.

In a similar fashion, Veremis describes Papandreou's foreign policies as "a balancing act between economic and security priorities, on the one hand, and domestic public opinion which has acquired a greater say in policy making than in the past, on the other" (Veremis 1988b, 265). It should be noted that Greece received substantial amounts of money from EU funds, about $ 800 million each year (Veremis 1988b, 264). Moreover, the basing agreement with the US brought rents amounting to $ 500 million a year (Veremis 1988a, 153).

Greek policies deviated from Alliance policies in many instances[81] in the late 1970s and early 1980s, Haass, in a similar fashion, concluded that "Greece therefore remain[ed] in the Alliance and continue[d] to provide the US with access to its facilities, less out of commitment to NATO's fundamental goals than through narrow self interest, based on the calculation that, vis-à-vis Turkey, Greece fares better with American bases and within NATO than it would without. What this adds up to is a difficult relationship in peacetime and an uncertain commitment and availability of Greece and its facilities in crisis" (1988, 64). Haass even suggested that the Alliance should prepare contingency plans for cases where Greek bases and forces would not be available. This would serve both as an insurance policy and as leverage against Greece (1988, 64). With regard to Papandreou's anti-US and anti-NATO rhetoric in contrast to his pragmatist policies, Loulis maintains that one factor making a real break with the West seem unprofitable was the notion that such a break would only benefit Turkey (Loulis 1984/85, 383). In NATO, Greece would be in a better position to attain the support of Western countries against Turkey (Veremis 1993, 187). Remaining in the Alliance, it could prevent the adoption of decisions which might put Turkey in a better position in the Aegean. McDonald maintains, too, that Papandreou had to abandon his pre-election call for the closure of US bases due to concerns over the balance of power with Turkey (1988, 81). If Greece left NATO, the sympathies of the US and NATO would tilt more towards Turkey (Coufoudakis 1992, 167). Besides, without the US and NATO, Greece could not modernize its forces which would put it in a disadvantaged position against Turkey. Thus, the US and NATO have remained important assets for Greece, assets that could pressure Turkey to alter its policies, or ultimately intervene to rescue Greece in the contingency of a Greek-Turkish war (Coufoudakis 1992, 152).[82] Nonetheless, as membership in

---

[81] Greece, for example, opposed the deployment of intermediate nuclear weapons as a response to Soviet SS-20 deployments and announced it opposed the continued stationing of short-range nuclear weapons on its soil. Greece, in fact, propagated the establishment of a nuclear-free zone in the Balkans (McDonald 1988, 82). Furthermore, Greece did not participate in sanctions imposed on Poland after the ban of the Solidarity trade union and the installation of martial law in 1982 (Veremis 1988a, 149). It accepted the Soviet explanation for shooting down a Korean airplane in 1983. Greece also criticized US attacks on Libya and did not implement EC sanctions imposed on that country. Moreover, Greece was criticized for its poor record in fighting terrorism (Haass 1988, 62).

However, given economic and military realities and the disappointment with Moscow's reaction to Greek attempts at improving ties given Moscow's desire not to alienate Ankara, Papandreou's foreign policy was less radical than his rhetoric in essence. In 1983, his government signed a new Defense and Economic Cooperation Agreement (DECA) with the USA (Ibid., 63-64). An agreement was signed in 1987 envisaging the purchase of 40 F-16s by Greece.

[82] Association with international institutions continued to constitute an important part of the Greek deterrence strategy against Turkey in the post-Cold War era. In the mid-1990s, Turkish engagement in the Balkans and Greek fears of encirclement, the Kardak crisis in the Aegean, and

NATO has entailed benefits unrelated to the Turkish threat and strategic realities would have necessitated membership in the alliance anyway, and given strong anti-NATO sentiments in the public, official rhetoric in Greece emphasizing the Turkish threat must, be seen, at least in part, as inward oriented and might have served to legitimize continued cooperation with NATO and the US despite rhetoric to the contrary.

Even though Turkey appears to have never tried to utilize the conflict with Greece in order to legitimize its membership in NATO, the disadvantages of remaining outside an institution where the opponent is a member must have been obvious to the Turks as well. Particularly the experience with the EU over the years highlights this point. When Greece was admitted in 1980, the EC stated that Greek membership would not influence the relations between the Community and Turkey (Stephanou & Tsardanides 1991, 210). Yet, this was a *promise* the Europeans could not live up to. On many occasions, Greece *captured* the institution to further its position against Turkey. For example, Greece blocked the reactivation of the association agreement between Turkey and the Community which had been frozen after the coup d'état in 1980 for some time. On another occasion, Greek approval of the customs union with Turkey could only be obtained after Greece was assured that the EU - in deviation from its original policy making membership conditional on a prior solution of the conflict - would start accession talks with Cyprus (Axt 1997, 259; Riemer 2000, 88). In the aftermath of the Imia/Kardak crisis, Greece blocked the release of EU financial aid, obviously in an attempt to punish the neighbor.[83] In 1998, France, Germany, Italy, and the Netherlands suggested that Cyprus should be admitted to the Union after a solution had been found (McDonald 2001, 133). Nonetheless, they were forced to give up this position when Greece threatened to block the entire project of enlargement. In the same year, Greece opposed the release of $ 410 million in EU-aid to Turkey citing "Turkey's provocative attitude towards Greece" as the

---

the S-300 crisis had contributed to a further deterioration of relations, and the Turkish threat had become ever larger in the eyes of the Greeks. One of the counter-measures to be taken by the Greek policy makers was to deepen the country's ties with institutions like the EU, WEU (Western European Union) or NATO. These moves rested on the assumption that Turkey would not dare to wage a war against a country so closely associated with the West (cf. Papacosma 2001, 211). Indeed, after the invasion of Cyprus by Turkish troops Greek politicians had undertaken measures to deepen the country's integration into European structures, such as the EC, hoping that EC membership would enhance the country's security vis-à-vis Turkey (Stephanou & Tsardanides 1991, 223; Meinardus 1982a, 445).

[83] Papahadjopoulos maintains that by blocking EU financial aid to Turkey in 1996, Greece simultaneously blocked the MEDA Financial Protocol and brought EU economic policy towards the Middle East and North Africa to a halt (1998, 56-57). In a similar manner, Greece blocked EU-FYROM relations in the period 1991-95 and prevented the release of aid to Albania in 1994.

reason (TDN, April 29th, 1998).[84] In general, Athens made its consent to Turkish membership in the EU contingent on an improvement of bilateral relations (Papacosma 2001, 212) and tried to refashion Greek-Turkish disputes as EU-Turkish ones. When Turkey was granted EU candidate status in 1999, it had to commit itself to resolving bilateral problems by 2004 and referring them to the ICJ if such a resolution has not occurred by the time. Turkey was also asked to facilitate the resolution of the conflict on Cyprus. In effect, the desire to enter into accession talks with the EU has been a major driving force behind the recent desire of Turkish policy makers to solve the issue as soon as possible.

Moreover, while it was usually Greece as the weaker side that tried to internationalize the conflict, and Turkey has never asked the Alliance to guarantee its borders against Greece, on some occasions it has captured the Alliance to further its position as well. For example, as mentioned before, Turkey wanted a revision of command and control responsibilities in the Aegean in order to lift its veto against Greece's return to NATO. Simultaneously, Turkish civilian governments used the veto against Greece reintegration into NATO as a trump card in their attempts to enter the EC (Güldemir 1986, 75, 103). On another occasion - though not directly related to Greece - Turkish PM Çiller threatened to bloc NATO enlargement if Turkey was not admitted to the EU. According to Bağcı, the Turkish Prime Minister was trying to divert attention from internal problems, and this policy was given up after the resignation of the Çiller government (1997, 585).

It was these experiences made with the creation of *inter-institutional* linkages and *capturing institutions* that rendered the conclusion of an agreement on the European Security and Defence Policy (ESDP) so difficult. Greece had joined the WEU hoping it would help to safeguard the country's security against Turkey (Papahadjopoulos 1998, 53). However, the WEU made it clear that the security guarantee could not be applied in cases where NATO members were involved. Moreover, Turkey was invited to become an associate member, enjoying most of the rights and obligations full-members were entrusted with. Nevertheless, with the autonomous ESDP of the European Union emerging, Turkey was about to lose its favorable position in the WEU, especially vis-à-vis Greece. The prospect of Greece utilizing the European Rapid Reaction Force against Turkish interests in the Aegean or on Cyprus was one of the major factors conditioning Turkish opposition against automatic access for the EU to NATO capabilities in cases where NATO as a whole was not engaged and the EU acted on its own. Turkey

---

[84] Greece was to change its stance after the 1999 earthquakes and lifted its veto in an atmosphere of détente.

lifted its veto only after agreement could be reached that Cyprus would not be able to participate in those EU operations making use of NATO assets and that the European force would not be used in the Aegean or on Cyprus.

However, despite the fact that a two-year-deadlock on the ESDP issue could be solved at last in Copenhagen, the formula found seems to cause after-pains, and disputes between Greek Cyprus and Turkey continue to impede EU-NATO cooperation in the process of the EU taking over the mission in Bosnia from the Atlantic Alliance. According to a report by the *International Herald Tribune*, the problem arises due to the fact that Cyprus and Malta have no clearance to share NATO secrets (cf. IHT, November 11th, 2004).[85] In practice, whenever the representatives of both institutions are to meet and tackle intelligence and security issues, these countries are asked not to attend the meeting. "The result is that the EU, on security issues, cannot function as a union of 25 countries," as Judy Dempsey, writing for the IHT, puts it. At the same time, there is some reluctance on behalf of NATO to share intelligence with the remaining EU member-states due to fears that this information might be transferred to the two countries in question despite the fact that they have no clearance. The daily cites EU and NATO diplomats stating that it was Turkey using EU enlargement to pursue its own agenda. On the one hand, Turkey is believed to be denying clearance to Cyprus (and Malta[86]) in order "to show its displeasure" over the Cypriot "no" to the UN plan this April. On the other hand, Turkey might be using the issue as a bargaining chip allowing it to pressure the EU to take a decision in favor of accession talks in December. Therefore a French diplomat is reported to have offered the view that Turkey might change its stance if the EU decides to start accession talks with the country or once the Cypriots opt for the UN plan.

Apart from the fact that NATO has been valued for reasons other than the cause of collective defense, as was the case during the debate on Greek reintegration into the NATO command structure, the membership of Greece and Turkey in the

---

[85] Simultaneously, on some occasions, even the representatives of Bulgaria, which acceded to NATO this year, are denied clearance to share NATO secrets given the lack of trust as some of the representatives were trained in the former Soviet Union (IHT, November 11th, 2004).

[86] Malta's case is somehow "unfortunate." Turkey does not object to the sharing of intelligence with Malta because of any bilateral disputes but simply because the same legal arguments Turkey is using in legitimizing its position vis-à-vis the Greek Cypriots apply to the Maltese as well. NATO rules require that only those countries that are members of NATO, or participate in the PfP program, or have any kind of a special security agreement with the institution are entitled to view NATO documents or share intelligence. Both Cyprus as well as Malta do not fulfill any of these criteria. However, Turkish officials have reportedly kept reiterating that they would back Malta if it applied for full-membership or at least decided to participate in the PfP again (cf. Radikal, November 16th, 2004).

Alliance has also led to the extension of their disputes into new areas, and problems related to the Alliance have caused additional tensions at times. For example, NATO infrastructure funds emerged as a new issue on the list of Greek-Turkish problems. Both countries raised objections to the projects to be funded by NATO on the opponent's territory, leaving both worse off than would have been the case otherwise. For example, in the period 1987-1988, Greece was to receive 8 percent of NATO's infrastructure funds and Turkey 14 (cf. Stearns 1992, 69). However, both countries raised objections to the funding of certain projects on the other's soil. Turkey rejected the funding of projects on the island of Lemnos, which in the view of the Turks had been illegally remilitarized by the Greeks. The Greeks, in turn, objected to the funding of projects located on the west coast of Turkey on the grounds the installations set up there could be used against Greece in case of bilateral hostilities. The objections raised by Greece and Turkey to the NATO projects resulted in Turkey receiving $ 252 million and Greece $ 144 million less than they were scheduled to receive (Ibid., 69).[87]

---

[87] In a similar manner, given their desire to secure a more advantageous treatment by Washington vis-à-vis the neighbor on the other side of the Aegean, both countries have been "willing to sacrifice tangible material gains for supposed positional advantages" (Stearns 1992, 46). According to Stearns, the US government was willing to make considerable concessions to Greece in an attempt to mend ties which were severely damaged given the relations US had entertained to the Junta and its failure to avert a Turkish invasion of Cyprus (Ibid., 47). Thus the terms of the Defense and Economic Cooperation Agreement (DECA) agreed on by Greece and the US in 1977 were much more than just favorable to Greece. Apart from the $ 700 million in aid Greece was to receive, in a letter to the Greek Foreign Minister Bitsios, the US Secretary of State Henry Kissinger granted some sort of a "hedged out but not insignificant" security guarantee, as Stearns puts it (Ibid., 47), and an assurance was given that Greece would receive certain scarce items it was interested in, including AWACs (etc.). Despite the fact that this DECA of 1977, in addition to the Kissinger letter, contained such favorable terms, Greece refrained from signing the agreement. Greece feared that once the agreement was signed, Congress would activate the 1976 DECA signed with Turkey, which would lead to the lifting of the arms embargo. Stearns maintains that "[l]obbying energetically behind the scenes against lifting the embargo, the Greeks delayed cashing in on their own agreement" (1992, 47-48).

Nonetheless, the Greek strategy did not materialize, and the embargo was lifted in 1978. This time the Turkish government refused to accept the 1976 agreement. It wanted to be compensated for the equipment they were entitled to have but did not receive in 1977 and 1978 due to the arms embargo. Furthermore, in Turkish eyes, the decision to provide $ 700 million to Greece implied recognition of the Greek position by the US according to which Turkey posed a threat to its security (Stearns 1992, 48). Therefore, Turkey wanted a renegotiation of the DECA. Turkey, in the end, succeeded in securing higher levels of aid and a commitment by the administration to try to secure Congressional approval for the new DECA signed in 1980.

The Greeks then signaled willingness to resume negotiations on the 1977 DECA (Stearns 1992, 48). By then, however, the circumstances had changed and the atmosphere was not as favorable to Greece as had been the case in 1977. Instead of granting more aid to Greece, the Americans declared that they did not feel bound by the terms of the 1977 agreement which Greece had refused to sign. The negotiations collapsed and were to be resumed with the PASOK government

The problems related to NATO infrastructure funds were partially solved during the short period of détente in the aftermath of the Davos meeting between the two prime ministers (Axt & Kramer 1990, 59). Greece agreed to lift its veto against the funding of a submarine monitoring station in the Turkish city of Çanakkale, while Turkey agreed to the funding of a port and an airport on the Greek island of Skyros. However, only a year later, NATO infrastructure funds were to cause troubles between Greece and Turkey again (Ibid., 65-66). Turkey vetoed the release of NATO funds for the modernization of a military installation on the Greek island of Rhodes claiming that a funding by NATO would mean accepting the militarization of the island. Although Turkey did not object to a national funding of the project, Greece, nevertheless, insisted on NATO funding and, in response to the Turkish veto against the Rhodes project, vetoed the Turkish chapter as a whole

Apart from that, in 1984 both countries vetoed each other's national force chapters. These chapters contain detailed information on the forces member countries have assigned to NATO. Greece had tried to assign forces on Lemnos to NATO (Axt & Kramer 1990, 30). As Turkey vetoed such an inclusion, Greece responded by vetoing the Turkish chapter, openly stating that this was done for political reasons (McDonald 1988, 77).[88] Such information is needed for NATO force planning and allowed for the comparison of allied forces with those of the Warsaw Pact during the Cold War years thence. Given the Greek-Turkish habit of vetoing each other's chapter, the annual document "Comparison of NATO and Warsaw Pact Forces" could not be issued after 1984 (Stearns 1992, 69). However, officials in Brussels asserted that the vetoing of each other's force chapters since 1984 did not constitute a major problem for the Alliance (Interviews in Brussels,

---

later in 1982.

The negotiations with PASOK were suspended in early 1983 when plans by the US administration to grant $ 755 million in aid to Turkey in the FY 1983/84 while intending to leave aid to Greece on the level of the previous year, that is $ 280 million, became public (cf. Meinardus 1982a, 479-481). The Greek American lobby started campaigning for a redistribution of aid. For instance, the Archbishop of the Greek-Orthodox Church of America, Iakovos, discussed the issue with Secretary of State Schulz and Secretary of Defense Weinberger. In the end, the Congress revised the levels of aid proposed by the administration on the basis of the 7:10 ratio leading to the initializing of the DECA in June 1983.

[88] According to McDonald, there are contingency plans including Lemnos (1988, 77). US troops are planned to land in Lemnos to assist in the defense of the Straits in case of hostilities.

Uzgel maintains that Greece brought forward certain reservations against a NATO decision related to the reproduction of chemical weapons (2001b, 76) and declared it would not withdraw its reservation until Turkey lifted its veto against the inclusion of forces on Lemnos. As NATO and the US assigned much relevance to the program, Turkey came under pressure to accept the Greek offer. In the end, Greece agreed to lift its reservation as an agreement had been reached with the US that their planes would be allowed to land on Lemnos in cases of hostilities.

July 2001). Both countries, nevertheless, provide NATO with the necessary information on their forces. The Alliance receives the Defense Planning Questionnaires completed by both countries and the review process takes place as usual up to the point where the force chapters are presented to the ministers. It is only at this level that Turkey blocks the assignment of forces on Lemnos and Greece in response vetoes the Turkish chapter as a whole. In the end, the member countries and the Alliance know about the capabilities of these countries, and the review process is carried out even though it lacks formal conclusion. The problem is more of a political and legal character and the repercussions on the defense review process of the Alliance are only minimal according to NATO officials.

Another problem within the context of NATO that has not been fully resolved yet is the issue of exercises in the Aegean Sea. Greece has usually desired a number of islands whose militarized status is contested by Turkey to be included in NATO maneuvers in the Aegean. On the other hand, NATO has chosen not to include these islands in maneuvers in order to avoid the impression it was taking side with one or the other party. However, this stance has been interpreted by Greece as pro-Turkish, and thus Greece has continuously refused to participate in exercises. On the other side of the Aegean, Turkey withdrew from NATO maneuvers in 1997 for the first time when some parts of the exercises to be held in September were put under the command of a Greek military officer to induce Greek participation after an absence of 17 years (Uzgel 2001c, 293). Indeed, some authors claim that instead of taking part in exercises aimed at training for the event of war against a common adversary, most of the exercises carried out by Greece and Turkey in the Aegean since 1974 have been directed against each other (see Stearns 1992, 70).

The situation was to change with the atmosphere of détente that emerged in 1999. In June 2000, Greece and Turkey finally once again participated together in NATO exercises The Turkish daily *Milliyet* proclaimed, "We have saved the Greeks," in regard to the fact that the maneuvers within the framework of the Dynamic Mix-2000 exercises envisaged the landing of allied forces, including Turkish troops, on Greek territory to fight against the invasion forces of a hostile country codenamed *Browland* (Milliyet, June 3rd, 2000). Unfortunately, problems related to the status of Aegean islands were to hamper NATO exercises once again only a couple of months later. The NATO exercise Destined Glory 2000 was more or less of historical relevance as Greek troops entered Turkish soil and Greek fighter jets landed in Turkey (cf. Demirbaş-Coşkun 2000, 27). However, after the exercise had begun, tensions arose between Greece and Turkey because Greek planes had been using the airspace over the islands of Lemnos and Ikaria whose militarization is one of the contentious issues between Greece and Turkey. In the end, Greece withdrew from the exercises before the scheduled end of the maneuvers.

The dispute related to the status of certain islets in the Aegean was also to have repercussions on the working of the Alliance. Some months after the incidents related to the Aegean islet of Imia/Kardak, NATO military officials gathered in Naples to prepare an exercise scheduled for September 1996. Greek officials then demanded that the island of Gavdos situated to the southwest of Crete be included in the maneuvers. Turkish officials, however, claimed the status of Gavdos was unclear and that the island could thus not be included in Alliance exercises (cf. Fırat 2001e, 469).[89] Greece complained to the allies and asked them to put pressure on Turkey to change its stance. Greek Foreign Minister Arsenis, who outlined the Greek view to his Canadian, British, French, and German counterparts as well as to Javier Solana during a NATO ministerial meeting in June 1996, told reporters that NATO would have to take important decisions in the following months on issues like the new command structure and the financing of regional headquarters. Drawing attention to the fact that these decisions required unanimity, he warned that the Greek stance would be dependent on the reaction of the Alliance towards "Turkish provocations" (ANA, June 15th, 1996). In the end, the issue was solved when the US declared that there had been a misunderstanding between the two countries and that Gavdos was undoubtedly Greek territory.

Sometimes, even "none-issues" seem to cause tensions between Greece and Turkey and hamper the work of the Alliance. According to the Turkish daily *Hürriyet*, a dispute between the two delegations over whether the strait of İstanbul should be referred to as *Bosphorus* or as the *Strait of Istanbul* had blocked the issuing of a number of NATO documents for months in 2002 (cf. Hürriyet, June 2nd, 2002).

Issue-linkages could be employed in situations where both parties are ready to negotiate, agree on the subject of negotiations, and are willing to make

---

[89] This statement by military officials caused some uneasiness in the Turkish Foreign Ministry which thought that the stance of the military was exaggerated, or as Fırat puts it, Gavdos was geographically a bit too far to assert such claims (cf. Fırat 2001e, 470). Even after the issue had been solved in the Alliance and the US declared there had been a misunderstanding, it caused further discontent between the foreign ministry and the general staff when a booklet produced by the general staff for internal use only became public causing uproar in Greece as it claimed all islets within the six-mile zone belonged to Turkey.

compromises. While extremely rare, this has been the case within the framework of the Greek-Turkish dispute on some occasions. The lifting of the Greek veto against the reactivation of the association agreement between the EU and Turkey in exchange for the annulment of a decree violating the rights of the Greek minority in Turkey constitutes a case in point. Notwithstanding the fact that the issue of minorities constitutes one of the problems dividing both countries, it is of minor relevance and compromises made in this area are not expected to prejudice the outcome of other disputes. Moreover, the reactivation of the association agreement could not be brought into a direct link with Greek sovereignty rights and was thus acceptable to Greece.

The aforementioned case of the mutual lifting of vetoes against certain projects funded by NATO on the soil of the other party shows as well that compromise is possible in areas which do not belong to the list of *tough issues*. While Turkey blocked the funding of projects on the island of Lemnos, it did not object to such funding on Skyros, whose status is not contended. Greece, in turn, accepted NATO funding for projects in Turkey. Even though this case does not correspond to the notion of issue-linkage, it shows that compromises can be made on issues not included on the *black list* or on those that do not prejudice the outcome of disputes related to other issues. This might even be the case during times of heightened tensions. For instance, at a time when Greece had been complaining about "an escalation in Turkish provocativeness" and blocking the EU's MEDA funds to prevent the release of aid to Turkey in mid-1996, both countries' telecommunication companies were reported to have agreed on the construction of an underground optical fiber network linking both countries to each other (ANA, May 30th, 1996). A total of 80 percent of the costs involved were to be carried by NATO. In fact, both countries have finally managed to sort out their differences related to the release of NATO infrastructure funds (Interviews in Brussels, July 2001).[90]

On the whole, in the Greek-Turkish case *issue-linkages* seem to be especially valued in terms of *blackmailing or retaliation* and are often established on an inter-institutional level. As both countries have not been able to work out an inventory of their problems, since Greece has usually claimed the only issue of

---

[90] Officials at NATO maintained that the problems related to the infrastructure funds had been solved a couple of years ago, though not further elaborating on the issue. During a telephone interview, an official, stating that he did not know exactly what kind of formula had been agreed upon, suggested that since Turkey had still not lifted its objections to the funding of projects on islands illegally militarized in the Turkish view, it would be plausible to assume that no further requests are being brought forward for the funding of projects on these islands.
Yet, it is also plausible to think that Turkey might have accepted the funding of civilian projects on these islands in return.

contention between the two countries is the delineation of the shelf while Turkey has been asserting that problems such as the breadth of Greek territorial waters and airspace, the status of certain islets and rock formations in the Aegean, or the militarization of a number of Aegean islets (etc.) require a resolution as well, it is hard to imagine how linkages could be employed to solve them. Since Greece talks about unilateral and *arbitrary* Turkish claims and asserts that Greek sovereign rights cannot be made the subject of negotiations, one cannot expect such issues to be made the subject of linkages facilitating a solution. This is why linkages established between contentious issues and issues not related to the conflict have usually proven destructive and why both countries have tended to isolate the first from the latter when they wanted to improve ties. This has also been the strategy both countries have been following ever since the rapprochement process started in 1999. They agreed to examine ways to facilitate cooperation in the fields of economy, trade, environment, tourism, and culture, etc (Papacosma 2001, 217). They expected cooperation on *unproblematic* issues to contribute to a better climate that might allow them to deal with the contentious issues dividing them thereafter.

5.3) Consultations and Transparency

The Greek-Turkish relationship within NATO seems to correspond to Patricia A. Weitsman's (1997) concept of *tethering alliances*, whereby states join or in this case remain in the alliance with the expectation that transparency and opportunities for consultations provided for by the institution will help to contain and manage their conflict (Weitsman 1997, 157). While such *tethering alliances* need not necessarily bring about a resettlement and give rise to amity between two adversaries, tethering can, at least, enable the states in question to lower the "probability or possibility of war" (Ibid., 165). Tethering can be facilitated by the existence of a third actor who enjoys good ties with both parties and acts as a mediator (Ibid, 164). Furthermore, the benefits of joining ranks with the threatening state might seem greater when there is an additional source of threat other than the ally (Ibid., 166). The alliance might not display a high degree of cohesion, but it is expected to prevent conflict by providing a certain degree of transparency and offering opportunities for consultations among members.

NATO, according to Weitsman, shows the elements of a tethering alliance since it has, in her view, contributed to the management of the German question and of Greek-Turkish disputes (1997, 190-191) [91]. The level of internal threats has

---

[91]In a similar fashion, Wallander and Keohane assert that old NATO had been a hybrid institution throughout most of its existence (1999, 41). Apart from its primary task of deterring the Soviet threat, that is to say fulfilling the functions of an alliance, it had to reassure members about

declined over time while NATO membership contributed positively to an improvement of relations among members. Notwithstanding the fact that the claim made here is that NATO membership has not only had positive effects on Greek-Turkish relations since issues related to NATO, such as command and control rights or NATO infrastructure funds and exercises, have added to the list of problems burdening Greek-Turkish relations, it is out of question that the Alliance has simultaneously served as an important and sometimes as the only forum for communication between the parties.

NATO has, on the one hand, served as a forum where formal deliberations or consultations could take place on Greek-Turkish issues. Especially during times of crisis when communication was essential but was not possible on a bilateral level, Turkish and Greek officials have kept in touch in NATO forums (see section on mediation). NATO has enabled the adversaries to uphold contact even when tensions reached the climax and war seemed imminent, as was the case during the 1974 crisis. Throughout this crisis, Greek and Turkish officials kept in touch with one another during NAC meetings and "dozens" of meetings in the

---

Germany and control internal security-dilemmas or mediate disputes among members thus fulfilling security-management tasks. In other words, NATO possessed specific assets for achieving transparency, integration, and negotiation among members, assets which were not related to the Soviet threat (Wallander 2000, 712, 716). The authors assert that the fact that NATO has always been a multi-purpose institution is one of the factors responsible for its continued existence.

In contrast, Haftendorn points to the example of the Warsaw Pact that disappeared even though it had also been a hybrid-institution (1997, 30). On the one hand, it fulfilled military tasks. On the other hand, for the Soviet Union it served as a means of control over its allies. She argues that the Warsaw Pact might have failed to survive despite the fact that it had been a multi-purpose institution because of underlying changes in the preferences of the member-states and because it was not recognized as a promising mechanism to solve the new cooperation and security problems. In fact, alliances generally seem to fulfill many tasks and constitute hybrid-institutions. Osgood lists four functions of alliances: pooling of resources to deter external threats, promoting the internal security or stability of members (for example, by legitimizing material assistance or military intervention against the opposition), restraining allies and securing their safety against each other, and establishing an international order (Osgood 1989, 458-461). According to Osgood, the internal concerns will become more prominent the longer the alliance endures and the more the external threat (or its perception) decreases. However, not all the functions of the alliance will be made explicit. Osgood maintains that "[i]n any case, the full substance and significance of an alliance is seldom revealed in the formal contract or treaty for military co-operation, any more than the essence of marriage is revealed in the marriage certificate" (Osgood 1989, 459).

With regard to NATO's collective security functions during the Cold War, Duffield maintains that those people offering pessimistic views about NATO's future underestimate its contributions to the quality of relations among members (1994-95, 767). In his view, NATO has made the use of force among members "virtually inconceivable."

NATO headquarters (NYT, July 23rd, 1974).[92] Apart from the multinational level, NATO ministerial meetings have often served as an opportunity for Greek and Turkish officials to tackle contentious issues on a bilateral level. Since these meetings are held regularly, they enable officials to meet behind the scenes without raising the expectations of the public, as would be the case otherwise. The fact that Greece did not want to be involved in bilateral dialogue on the contentious issues until only recently added to the value of such *informal* meetings, since they helped to avoid the impression Greece had been drawn into a dialogue with Turkey. The value of Greek-Turkish gatherings on the sidelines of NATO seasonal meetings is also acknowledged by Coufoudakis, who maintains that "[e]ven though these meetings failed to actually resolve any of the differences between the two countries, they provided an important means of communication through which mutual positions were explored and misperceptions were possibly clarified" (1985, 208).

Even though they failed to facilitate a final resolution of the conflict, permanent contacts in NATO headquarters as well as meetings on the sidelines of NATO ministerial gatherings or summits have, nonetheless, brought about valuable results on many occasions. Writing on the crisis of 1964, Windsor maintains that "Paris thus gave opportunities for the kind of informal diplomacy and corridor conversations which could not take place anywhere else, and also a means of putting pressure, in the last resort, on either Greece or Turkey, or both, should they impede any agreements. It may be that the distinct improvement in the situation in Cyprus after the Turkish air raids at the beginning of August was due to such a combination of functions in the NATO building in Paris" (Windsor 1964, 15). Later on, when the first phase of inter-communal talks between Greek and Turkish Cypriots ended in a deadlock in 1971, the representatives of Greece and Turkey discussed the issue on the sidelines of the NATO foreign ministers meeting of May 1971, and these *corridor meetings*, as Stearns deems them, brought about an agreement that the inter-communal talks should continue (Stearns 1992, 113-114). The Madrid Declaration of 1997, regarded as a "non-aggression pact," was also signed on the sidelines of a NATO summit. The

---

[92] Apart from the deliberations in the Council, NATO's direct communication lines with the member countries also proved useful (cf. Tülümen 1998, 150). As a result of the crisis, the telephone lines in and to Ankara were blocked. Thus, foreign officials had to make use of NATO lines when they wanted to contact the officials in Turkey. For example, the British Foreign Minister Callaghan sent messages intended for his Turkish counter-part to the UK delegation in NATO. The members of the delegation, thereupon, informed their Turkish colleagues about the messages. The Turkish delegation, in turn, would transmit the messages to the Turkish Foreign Ministry. Sometimes messages back followed the same path.

importance of NATO with regard to communication was also acknowledged by the officials of both countries. In October 1998, the Greek and Turkish prime ministers "concurred that, if their respective countries had any significant channel for communication, among their wide range of differences, it was at the military level within the ranks of NATO" (Papacosma 2001, 214).

Maybe what renders this channel so attractive is the fact that it enables for continuous and, above all, *informal* contacts. Of course, this is a general asset valued by all members. Such informal encounters are, in the view of Kay, "the most effective from of consultation that developed in NATO (Kay 1998, 38)". Kay, too, points to the importance of "hallway negotiations" which enable representatives to exchange views or to gain the others' support for a particular initiative in the NAC (cf. Ibid., 38-39). Despite the difficulty to determine the impact of such negotiations on the policies and strategies of the member-states given the informal and confidential character of such discussions, the fact that the representatives have chosen to continue a tradition to meet during a luncheon prior to the meeting of the NAC, which emerged in the early 1960s, shows that the member-states see certain benefits in such informal meetings. Such luncheons enable the representatives of the member-states and the secretary-general[93] to deliberate on international developments in a relaxed atmosphere, which might even move a representative to ask his country to change its stance to facilitate consensus - of course, as such a revision of position need not always occur and the representatives are bound to the instructions they receive from their countries, a consensus arrived at during a luncheon does not guarantee that an agreement will subsequently result in the NAC (Ibid., 39; for these meetings see also Jordan 1979, 184).[94] Nonetheless, the representatives have not stopped valuing these meetings or called them into question as a whole (see Kay 1998, 39):

> By the 1990s, the weekly lunch meeting had become deeply ingrained in the institutional culture of NATO. It is held at the private residence of each NATO ambassador as well as those of the secretary general and deputy secretary general on a rotating basis.

---

[93] In the initial phase, Secretary-General Stikker refused to participate in these meetings. In fact, this tradition to meet informally to discuss issues of relevance was established by small countries that had difficulties in gaining access to the Secretary-General during Stikker's tenure (cf. Jordan 1979, 126). As such meetings allowing the participants to work out positions to be adopted in the next Council meeting proved their usefulness, most of the other member countries joined this process initiated by the small powers.

[94] During the tenure of Manlio Brosio, the Secretary-General, the Chairman of the Military Committee, the SACEUR, the Assistant Secretary-General for Defense Planning, and a few others also started to gather for such informal luncheons which were to become institutionalized (Jordan 1979, 231).

The agenda is usually set by the secretary general or his deputy as well as by the host country and the remaining ambassadors as required. Generally, the substance of the next day's formal ambassadors' meeting in the NAC will be discussed and national positions made informally known to the others at the meeting. This gives the ambassadors an opportunity to correspond with their home governments and either confirm or seek changes in government policy in advance of stating their formal position in the NAC. It also gives the secretary general an opportunity to raise issues to the national delegates on behalf of the organization and explain how particular decisions might affect NATO as an institution. Informally, the lunch also serves as an opportunity for ambassadors to blow off steam around their colleagues in a less formal manner than that provided by the NAC if there are serious disagreements within NATO. The lunch thus continues to be one of the most important informal patterns of consultative behavior within the state-dominated process of NATO decision-making. Indeed, to a certain degree the informal lunch process diminishes the importance of the official NAC meetings themselves as actual business occurs the day before.

Even though there is no concrete information proving that Greek and Turkish representatives or the other ambassadors have used these luncheons to tackle differences between the two Aegean neighbors, it is very plausible to assume that Greek-Turkish problems might have constituted the topic of such informal meetings on one or another occasion.

With regard to transparency, NATO procedures seem to allow member-states a deep insight into the capabilities of others. The Defence Planning Questionnaires (DPQ) provide detailed information on the arsenals of the other members and partner countries (Interviews in Brussels, July 2001). Still, there are a number of factors curtailing the value of NATO's review procedures. It goes without saying that even allies will have certain secrets and will withhold information regarded as being sensitive.[95] Moreover, NATO has to rely on the trustworthiness of the

---

[95] For example, Stearns notes that "[u]ntil the issue [of the militarization of islands] became academic in 1983, when the Papandreou government announced that Greece would no longer participate in allied exercises in the Aegean that excluded the island of Lemnos, which the Greeks had militarized over Turkish objections, it was easier to compose an Aegean scenario realistically projecting the capabilities of Warsaw Pact forces in the area than those of Greece and Turkey-which were kept secret from each other and from the alliance" (Stearns 1992, 72).

member-states when evaluating the data they have provided. There is no NATO intelligence service monitoring or collecting information on the military capabilities of member-states. One can assume that this is done by each state on its own, and these states will not always be willing to share the data with NATO. Güven Erkaya, the former Chief of the Turkish Sea Forces, who was once tasked with filling in the DPQ while an officer at the General Staff, remembers that the data provided in the DPQ did not necessarily reflect the real capabilities. The questionnaires were completed in a manner to secure the largest possible amount in aid as possible (cf. Baytok 2001, 16). Erkaya adds that NATO did not only know about this, but in fact encouraged such behavior. However, despite these drawbacks, NATO continues to guarantee a certain degree of transparency which might not be possible on a bilateral level, especially in cases where two states are in conflict.

Whether such a degree of transparency can help to avoid conflict or even facilitate cooperation is, of course, another question. Since Greece and Turkey perceive each other as threats to their security, they will probably scrutinize the data provided by the other party more carefully. Knowledge acquired on the capabilities of the other party may mitigate concerns as well as exacerbate mutual anxieties. An official at NATO maintained that while the DPQs provided detailed information, such information might indeed add to the anxieties of one of the parties or might even render plans for war easier. The official concluded, "as long as they see each other as an opponent, transparency does not count much" (Interviews in Brussels, July 2001).

Even so, the fact that parties might not be able to infer intentions from capabilities, as argued by Krebs (1999, 355), or that they might conclude the other party is harboring aggressive designs adds to the importance of NATO. While transparency on capabilities might give rise to concerns, consultation mechanisms might enable members to inform each other about their intentions or to clarify misperceptions. Of course, when two states are in conflict, a country might want to employ its military arsenal to send certain signals to the opponent, to communicate certain warnings, or even to intimidate. Thus, it will be up to the feuding allies whether they will use institutional forums for reassuring their allies or for showing their muscles and communicating certain warnings.

5.4) Military Assistance

Krebs as well as many Greek authors criticize the US and NATO for having created a military imbalance in the region favoring Turkey given its greater strategic importance. The military assistance provided to both countries is said to have exacerbated the intra-alliance security dilemma,[96] added to bilateral tensions, and made compromises less likely. What is more, in the view of some authors, the imbalance created was not an unintended side-effect but was actively sought by the US in order to put Greece in a position where its capabilities would not allow the country to wage a war against Turkey when tensions reached breaking-point.

Platias, for instance, argues that NATO and the US dominated Greek defense planning to an extent which "froze any at all Greek initiatives" (Platias 2001, 97), especially in the 1950s. The task the Greek forces were assigned to fulfill by NATO was to secure internal order while the defense of territory against external forces was only a secondary issue. Greek forces were expected to cause some delay against the aggressor until reinforcements could arrive (see also Veremis 1988b, 252). This primacy of internal mission, according to Platias, left its imprint on the force structure of the Greek army. As a consequence, Greece lacked strong naval and air components and was in no position to defend itself "autonomously" against external attack (see also Roubatis 1979).

Roubatis goes a step further and asserts that the US actively sought to constrain the capabilities of Greece. The crises of the 1960s between Greece and Turkey had shown to the US that a war between Greece and Turkey was more likely than Soviet bloc aggression against Greece (1979, 49). Thus, the US was interested in a restructuring of Greek forces in a manner that would prevent Greece from fighting a war against Turkey. They pressed for reductions of Greek air and naval capabilities and refused to furnish Greece with military fighter jets as demanded. The junta was to accept most of the suggestions made by the Americans (Ibid., 51). Yet, US tactics to keep Greek naval and air capabilities at a low level continued in the aftermath of the junta years. However, a new state of mind had emerged in Greece. The crisis of 1974 had shown how dangerous this strategy was for Greece vis-à-vis Turkey. From now on, Greece was to put more emphasis on developing its own capabilities and try to develop an autonomous strategy. Greek naval and air forces were strengthened, certain forces were left outside NATO's integrated structure, and those forces assigned to NATO were to be put under national command as soon as Turkey undertook any actions threatening

---

[96] Note that the CFE Treaty of 1990 entailed a flow of military material from NATO countries in the center to Greece and Turkey, since the provisions of the treaty envisaged a reduction of forces in Central Europe while allowing Greece and Turkey to build-up their arsenals in order to preserve the balance between the two former blocs (Riedel 1996, 18). According to Riedel, the cascading program of NATO entailed increases in the military arsenals of Greece and Turkey by 18 and 24 percent respectively in the period 1990-1995. Moreover, as mentioned elsewhere, almost half of the Turkish territory was exempted from the jurisdiction of the treaty.

Greek interests (Platias 2001, 98). Furthermore, the arms industry would be strengthened to lessen dependence on external arms supplies.

This line of argumentation, however, seems to be *inaccurate* and *exaggerated* in certain instances. Firstly, the argument according to which the sole function of Greek forces had been to uphold internal order does not seem to be fully correct. For instance, Rupp argues that Greece's role had not been confined to maintaining internal order. Greek forces had been tasked with defending their homeland, securing lines of communication, and participating in NATO's naval operations in the Eastern Mediterranean alike (Rupp 1988, 30). This view is also shared by Iatrides, who claims the argument that Greek forces had solely been equipped for internal tasks is not totally correct, due to the fact that "[t]he naval and air forces had been gradually given important assignments beyond Greek space" (Iatrides 1993, 20).

Secondly, in contrast to the claim according to which a military imbalance was created in the region due to the arms shipments in the face of the requirements of the Atlantic Alliance, when comparing the aid levels Stearns identifies an *inconsistency* in US aid policies and holds that aid levels "have been influenced by factors other than Greece's and Turkey's "particular NATO-related requirements" (1992, 45). During the period of the Civil War and immediately afterwards Greece received aid in a ratio of two or three to one (Ibid., 45). Between 1952 and 1964, Turkey received much more aid than Greece, except for 1953, where the ratio only slightly favored Turkey (8:10), and for 1960, where the ratio was tilted in favor of Greece (11:10). In the view of Stearns, these exceptions did not occur due to any "obvious military reason" (Ibid., 45). In the years 1964, 1965, and 1966 the ratio was almost balanced and Turkey received only slightly more than Greece did. In the period 1967-1974, the ratio was approximately 4:10 in favor of Turkey. This was to change after an arms embargo was imposed on Turkey. Turkey received no military assistance in 1976, and in 1977, the ratio was 12:10, whereas in 1978 and 1979 the aid levels were fairly balanced (10:10).[97] From 1980 onwards, Congress began to disperse aid in a ratio of 7:10 (except for the FY 1992).

An official at NATO maintained that it is possible that NATO might have concluded that Turkey needed more forces than Greece given its location and

---

[97] Turkey received aid in the fiscal years 1976-1979 even though an embargo was imposed as "pipeline" aid continued as well as other forms of assistance not requiring new appropriations (Stearns 1992, 167, footnote 10; see also Sönmezoğlu 1995, 95).

Note that during the embargo years, Turkey received additional aid from Germany worth DM 50 million (Pleninger 2002, 123).

functions (Interviews in Brussels, July 2001). The apportionments made by NATO were based on the Soviet threat, and Greek-Turkish disputes could not be allowed to interfere. Moreover, when such apportionments were made, the Alliance also asked the question of feasibility. Thus, factors like the GDP of the country in question were also taken into account. Finally, even if Turkey had been apportioned more material or forces, this would not have foreclosed the possibility for Greece to acquire additional forces in order to guarantee its security vis-à-vis Turkey. Greece would have always had the possibility to *remedy* these deficits. Another official maintained that NATO allowed for a certain degree of coordination, but it was ultimately up to the members what they wanted to do. The member-countries could cooperate to the extent they wanted to. NATO was in no position to enforce anything. For instance, France had opted out of the military structure because it had wished to do so. Greece had chosen to remain outside the command structure for some time, too. The official also cited the case of Germany whose contribution has usually been lower than its financial capacities would allow for.

The point made here is that the primary responsibility for the size, equipment, organization, maintenance, and funding of their military forces lies with the member states in spite of the fact that there is some degree of coordination and that the national policies are influenced by the existence of a command structure as well as by the requirements of NATO defense planning (see NATO 1999, 203; International Institute for Strategic Studies 1998; 38). In the end, it would have been up to Greece to strengthen its capabilities to counter the perceived or actual Turkish threat.

Another valid point is that it is hard to conceive why a military alliance "outgunned and outmanned," as McDonald puts it (1988, 75), by the opponent should try to limit the military capabilities of member-states. Stearns, holding that NATO plans throughout the Cold War did not envisage many contingencies for joint operations by the Greek and Turkish forces and that both countries' main mission was to defend their homelands until enforcements arrived, asserts that it was due to this "coincidence of their national and NATO military missions" that significant adjustments in their force deployments and defense plans were not directly challenged by the Defence Planning and Military Committees. NATO chose to treat forces directed against the other NATO partner as units fulfilling national defense requirements consistent with NATO plans" (Stearns 1992, 73-74). To put it differently, there is no reason why NATO should have objected to a development of Greek military capabilities, since this would have added to the capabilities and the deterrence of the Alliance as a whole. In fact, Greece, just like Turkey and Portugal, failed to fulfill NATO force goals in the period 1970-1983 in spite of the fact that its defense expenditures had increased at a faster rate than its economy and thus imposed a strain on its economic development (Rupp 1988, 28).

If Greece had not developed the sufficient capabilities to respond to the Turkish threat due to the constraints imposed by the US in particular, or by NATO in general, the question arises as to why this was to change in the post-1974 era. One could argue that Greece was outside the military structure during this period. However, Greek cooperation with the US continued and, if the argument that US policies had prevented Greece from acquiring the necessary capabilities was accurate, the US would have had the opportunity to do so during this period as well. Greece also continued its military build-up and the modernization of its forces irrespective of the fact that it had returned to NATO's military structure in 1980. If the constraints of membership had been so decisive in the pre-1974 period, why have the same constraints not prevented Greece from becoming the member-state in NATO spending the highest percentage of its GDP on defense requirements thereafter?

In fact, Stivacthis' description of Greek defense behavior seems to be more convincing. He claims that Turkey, maintaining a large standing army and drawing on huge economic resources, has put more emphasis on internal rather than on external balancing (2001, 85-86). This trend was reinforced after the end of the Cold War. Greece, on the other hand, relied on external balancing both for deterrence as well as defense. The recurring crisis on Cyprus and the 1974 invasion of the island by Turkey showed the shortcomings of this strategy. Hence, from 1974 onwards, Greece started to pay more attention to developing its own resources and modernizing its forces while limiting the constraints imposed by NATO membership (Stivachtis 2001, 87-88). Thus forces in Attika and on the islands were not assigned to NATO (McDonald 1988, 73). Furthermore, a number of airfields were constructed on certain Aegean islands in the late 1970s and early 1980s, and new submarines, fast patrol boats, and new frigates were purchased in order to strengthen the navy (Ibid., 74). Today, Greece is the NATO country spending the largest percentage of its GDP on defense and allotting the highest portion of its labor force to defense (Stivachtis 2001, 92).

Yet, the often made assertion that participation in NATO's integrated command structure and force planning constrained the sovereignty of the member-states and their ability to make use of their national forces for individual purposes seems to be inaccurate as well. A NATO official referred to this line of argumentation as a *myth*, especially common in France (Interviews in Brussels, July 2001). Another official maintained that the assignment of forces to NATO had no implications in practice, since the transfer of authority took place only in crisis situations. In peacetime, member-states retain full control over their forces and are free to decide where they want to deploy these forces or for what purposes they should be used. For example, NATO's command structure had not prevented Turkey from using airplanes assigned to NATO for air raids against Cyprus in 1964 or to mass troops along the Syrian border and threaten war. Thus, the argument according to which Greece did not assign some forces to NATO in order to be able to deal with the Turkish threat more effectively in crisis times makes sense only so far as it may be a strategy for showing concern over the behavior of an ally and appeasing domestic anti-NATO sentiments. Indeed, one could argue that it was in this connection that NATO made its most important contribution during the 1974 crisis. Leaving NATO's integrated command structure served as an "exit strategy," or to put it in other words, as an alternative to waging war against Turkey for the Karamanlis government. This is not to say that there would have been a war had it not been for NATO, but that Greek withdrawal from the military arm of NATO enabled Greek policy-makers to appease public sentiments to a certain extent and escape the impression of having done nothing in the face of the Turkish move on Cyprus, which would have bolstered the position of the Colonels.[98]

The question also arises as to why the 7:10 ratio, which is arbitrary in nature, is viewed as a measure to preserve the balance of power. It is hard to understand how this balance is defined. Turkey is much bigger than Greece with regard to geographic size and population. Consequently, its standing army is larger than the Greek forces almost by a rate of 3. Given these numbers, it seems to be natural that Turkey was granted more aid, and must indeed have been in need of more aid.

Moreover, the question arises as to what would have been the case had these countries not been NATO members. As an official put it, Turkey would have had to acquire huge capabilities anyway given its geographic location and threat perceptions (Interviews in Brussels, July 2001). It would not be wrong to assume the same for Greece. While membership in the Alliance made the continuation of US aid more certain, it was not a precondition for it. They would probably have received aid even if they had remained outside NATO if US interests had necessitated this. It is worth mentioning that Israel and Egypt - two countries that

---

[98] In a similar manner, Turkish policy not to assign its Fourth Army to the command structure could be interpreted as a result of the desire to enhance the credibility of its threat of force against Greece should the latter try to change the status quo in the Aegean Sea (Interviews in Brussels, July 2001).

do not have institutional ties to the US within the framework of a military alliance as Greece or Turkey do - have been the largest recipients of US military assistance since 1979 (Stearns 1992, 43). Perhaps outside NATO such capabilities would have caused more concern. There would not have been the degree of transparency provided by NATO procedures. Moreover, they might not have been able to keep contact and engage in consultations as NATO has made it possible. As mentioned elsewhere, the issue of US aid levels sometimes caused friction in their bilateral relations, as well as between these countries and the US. However, it is hard to imagine that similar problems would not have occurred had these countries not been NATO members.

5.5) The Experience of Democracy in Greece and Turkey

As mentioned above, there are different views concerning the question of whether NATO made major contributions to the democratization processes in the cases of the members admitted in 1999, namely Hungary, Poland, and the Czech Republic. Whereas it might be difficult to say to what extent the prospect of membership or their association with the Alliance through the PfP facilitated the consolidation of democracy in these countries, in the cases of Greece and Turkey, it is possible to look at the course of their democratic development as well as at the current state of democracy and draw conclusions concerning the question as to whether 52 years of membership in the Atlantic Alliance have made a difference. While Reiter's assessments presented in section three have already provided for a first depiction of the situation, this section provides a more detailed description and discussion of the situation in these countries.

First of all, the aspect of democracy seems to have played a role in the calculations of Greece and Turkey when they joined NATO. One of the reasons why the Democrat Party in Turkey sought membership in NATO was the expectation that NATO membership would foreclose any attempts to remove the government from power forcefully. In the view of Ülman and Sander, the Democrat Party, which was in the opposition until 1950, regarded Turkey's membership in NATO as a necessary measure to guarantee the maintenance of the democratic system and to safeguard its own continued existence (cf. 1972, 6-7). The members of the Democrat Party (DP) were familiar with the fact that past attempts to move towards a multiparty system in the country had failed and feared that the İnönü government might say farewell to democracy once it recognized it would lose its grasp at power. They hoped that joining NATO, which they perceived as the "common front line" of Western democracies, would prevent the

İnönü government from playing such a game on the opposition.[99] As a consequence, during the election campaign of 1950 the Democrat Party focused on the domestic dimension of NATO membership rather than focusing on the security perspective and accused the Republican People's Party of leaving the country outside the "Western democratic front." As soon as the DP won the elections and assumed office in 1950, the realization of membership became an issue of the highest priority. A couple of days after assuming office, the DP government was ready to send troops to Korea to facilitate the country's chances to become a NATO member. Since the government viewed NATO membership as a necessary measure for its secured hold on power and turned its commitment to the West into a foreign policy principle, the fact that the Soviet Union revised its policy towards Turkey in 1953 did not entail any changes in the foreign policy choices of Turkey (Ibid., 7). The Alliance was perceived in a similar way in Greece that pressed for membership in NATO, among others, because it viewed the Alliance as "a door to a community of democratic European states" (Veremis 1988b, 242)"

In the case of Turkey, there is also evidence that the desire to obtain US military aid and later to join NATO moved officials to undertake efforts to *mend*, or, at the least, to touch up democratic shortcomings. Ülman and Sander assert that the desire of the Turkish government to win US support vis-à-vis the Soviet Union might have contributed to an abandonment of the single-party system in favor of a multiparty system (cf. Ülman & Sander 1972, 4-5). During debates on the US commitment to Greece and Turkey in Congress and in the public, the main argument against such a commitment was based on the fact that both countries did not really constitute democracies. In effect, one could not defend the "free world" through an extension of help to such countries. Responding to this criticism, President Truman kept repeating that both countries were of enormous strategic relevance and were making considerable efforts to fully democratize. In the view of the authors, such debates might have induced the Turkish government, which wanted to guarantee its security, promote its ties with the West, and receive military and economic aid at all costs, to accelerate the country's transition to a

---

[99] In 1959, the DP government and the US were to sign an agreement which called for mutual assistance in case of direct and indirect attacks (cf. Ibid., 9-10). In the view of the opposition, Turkey had already been under the US security umbrella and this new agreement was superfluous with regard to external threats. They believed that the term "indirect" referred to domestic developments and allowed the US to intervene should the DP government be removed from power with a coup d'état (Ibid., 9). Yet, when the military seized power in 1960 - according to Karpat an action "chiefly designed to answer a threat (if there actually was one) to the Republican People's Party (RPP) which had governed Turkey from 1923 to 1950 (1988, 137)" - the US did not intervene to help the Democrat Party back to power. The US did not prevent a military court from hanging the Prime Minister removed from power either.

multiparty system. In a similar fashion, Erhan points out that the military assistance agreement concluded with the US foresaw the subordination of all military affairs to a single government (2001a, 537). This had been one of the main reasons why the army had been put under the full responsibility of the government and all security forces were connected to the Ministry of Defense. Moreover, İnönü had resigned from the chairmanship of the RPP in order to conform to the idea that the president had to be neutral.

While there are no indications that the prospect of NATO membership caused Greek officials to refurbish the state of democracy, Greece undertook certain measures in order to promote compliance with NATO principles especially to promote the civilian control of the armed forces after its admission to NATO (cf. Kay 1998, 51). For this purpose, officers were sent to the NATO Defense College in Rome where they were taught how civil-military relations should look in a democratic regime. Moreover, in the period 1950-1961, more than 11,000 Greek officers were trained in the US (Veremis 1988b, 252). It was thus almost an irony of fate when Greek officers seized power in 1967[100] using a NATO counter-insurgency plan called Prometheus, which had, in fact, been devised for the contingency of internal subversion in Greece (Kay 1998, 51; Uslu 2000, 202).[101]

Kay points out that NATO, while facing a "crisis of conscience," did not do anything to discourage the Colonels in Greece. Whereas the Scandinavian members tried to raise the issue in the NAC, their efforts were resisted by the "more powerful members" and by Secretary-General Manlio Brosio. Brosio supported the view that if the security of the Alliance was weakened for the lack of freedom in a member-state, this might put the freedom of all member-countries in danger (Jordan 1979, 184). According to Uslu, more than half of the member-countries were severely criticizing the Colonels' regime and questioning its membership in the Alliance (2000, 240). It was the US that sided with the regime in Greece. While Greece was facing growing pressure in the international arena, high-level visits like that of the US Vice-President Spiro Agnew to Greece signaled the continuation of good relations between the two countries (see Clog 2002, 162), and US, UK and NATO military officials continued to meet the

---

[100] After tensions between the government and the palace led to Papandreou's resignation as the prime minister, new elections were scheduled for May 1967 (Clogg 2002, 157-160). However, before the elections could take place, a number of officers staged a coup d'état, which was, according to Clogg, "an attempt to pre-empt an almost certain Centre Union victory at the polls" (2002, 159). The right-wing officers expected Papandreou, whom they were suspicious of for his leftist credentials, to be the main figure in a Center Union government. In such a case, Papandreou, they feared, would probably purge the ultra right wing officers from the army.

[101] Note that the leader of the junta, Papadopoulos, had served as a liaison officer between the CIA and the Greek secret service KYP (Uslu 2000, 203).

Colonels on a regular basis (Kay 1998, 51). Furthermore, the Greek junta continued to receive economic[102] and military aid from the US despite the embargo imposed by the Congress on heavy weapons. The embargo was to be lifted as a whole a short time after it had been imposed (Veremis 1988b, 252). Strategic interests once more outweighed the allies' commitment to proclaimed community values such as democracy. The new regime under the rule of the Colonels was, meanwhile, eager to adopt policies favorable to NATO and the US and reaffirm its anti-communist stance (see also Varvaroussis 1979, 71; Uslu 2000, 240-241). For example, the US enjoyed far-reaching freedoms in the usage of the bases in Greece and a homeport agreement was signed in 1973.

As was the case with Greece, American policy makers had no hesitations in continuing their cooperation with military regimes in Turkey. Turkish military officials, in the meantime, usually tended to reaffirm their commitment to the West and to continue cooperation in the same manner as before. Hence, one of the primary concerns of the Turkish officers when they seized power in 1960 was that the US might stop providing economic and military aid to the country. Thus on the same day they seized power, they sent a delegation to the US embassy to assure the US of their continued commitment to NATO and CENTO[103] (cf. Harris 1972, 86-82). The US was quick to recognize the new regime and US President Eisenhower informed the new rulers in a letter that he was happy about the announcement that the country's ties with the aforementioned organizations would be preserved. So, instead of imposing any sanctions, the US preferred to increase the levels of military and economic aid to Turkey (Uslu 2000, 18). In this period Turkey and the US signed new agreements, and the regime in Turkey made amendments to the constitution rendering the process of getting parliamentary approval for the agreements signed with the US easier (Uslu 2000, 352).

One of the objectives of the new regime was to reduce the number of senior army officers in an attempt "to end the 'inflation of ranks' and to eliminate those whom they judged to be lukewarm about the new regime" (Harris 1972, 89). In order to avoid grievances, the officers who were going to retire were in line to receive generous payments. The military regime also did not want to take such steps without informing NATO. Therefore, SACEUR General Norstad was invited to Turkey to discuss the country's role within the framework of NATO's defense system and to ask him for assistance to make the retirements possible. General Norstad did not see any reason to object to the retirements as there were too many senior officers in the Turkish army. Moreover, Norstad told the Turkish regime that he would try to assist in securing the required financial means.

---

[102] For instance, the US provided the junta with credits on very favorable terms (Uslu 2000, 242).
[103] Central Treaty Organization

The 1971-1973 technocrats' government installed by the military was - like the regime of 1960 - inclined to adopt policies which would be appreciated by the Americans (cf. Uslu 2000, 24). In previous years, the issue of poppy cultivation had been causing tensions between the two allies, and former Turkish governments had refused to impose a ban. This was to change after the military intervention of 1971 when the new government prohibited poppy growing. The US, in return, offered aid worth $ 35 million to be used for the compensation and reorientation of the farmers (Harris 1972, 197). Simultaneously, leftist groups were pursued and suppressed with more vigor in order to win the favor of the US, while anti-American articles largely disappeared from the press (Uslu 2000, 355).

Strategic interests rather than concerns over the maintenance of democracy characterized US policy towards Turkey when the Turkish military seized power on September 12th, 1980 for the third time in 20 years. Looking back on the situation in and around Turkey at that time, Jimmy Carter stated in an interview with the Turkish journalist Güldemir that he had viewed the military coup of 1982 as a first step towards the stabilization of the region (Güldemir 1986, 95).[104] The factors rendering the US response so *sympathetic* towards the military takeover were manifold.

The Soviet Union had invaded Afghanistan in 1979 and the Shah's regime in Iran had been overthrown (cf. Güldemir, 1986, 65-69). In this context the Americans were concerned that with its international image damaged anyway, the Soviet Union might try to push through Iran downwards to the Persian Gulf. As Greece was still outside the military structure of NATO and Iran had left the *pax-America*, Turkey would be on its own in the event of such a Soviet military move. Apart from these international developments, internal turmoil and growing Islamic and anti-Semitic tendencies in Turkey had further added to US concerns. Moreover, Turkish policy makers had adopted an intransigent position towards US demands. For example, the Turkish Prime Minister, Süleyman Demirel, had insisted on a provision in the new Defense Cooperation Agreement limiting the use of bases to NATO contingencies. Thus, the US could not be absolutely sure whether they could use the base in İncirlik during a crisis in the Middle East. Demirel had also refused to agree to Greece's reintegration into NATO. However, a reintegration of Greece had to take place prior to the next elections in Greece, since PASOK, which had declared Greece might leave NATO should it come to

---

[104] In the 1970s, the Turkish political system was plagued with instability, and the country experienced 12 minority or coalition governments in the period preceding the coup (cf. Kuniholm 1983, 428). Moreover, open fighting had broken out between leftists and rightists leaving about 28 people dead everyday.

power, might win the elections (see Uzgel 2001b, 40).[105] For all these reasons, the impression arose that the US would condone any measure that would promote stability in the region and allow for a reintegration of Greece (Ibid., 71). Uzgel maintains that even if the role played by the US during the developments of September 12th might be open to speculation, what seems to be certain is that this coup was to the great benefit of US strategic considerations and that US-Turkish relations in the aftermath of the takeover could be described as almost *perfect* (Uzgel 2001b, 38).[106]

According to Güldemir, the Chief of the Turkish General Staff, Kenan Evren, told SACEUR General Rogers that *they* would agree to Greece's reintegration if they had the authority to do so a short time before the military had staged the coup (Ibid., 76). While later Kenan Evren was to claim such a conversation had never taken place, oddly enough, already on the same day the military seized power in Turkey, the Greek Foreign Minister, Mitsotakis, had told reporters that Greece would now be able to return to the military arm of NATO (Ibid., 76). [107]

---

[105] On October 20th, 1980, the *New York Times* reported that former PM Demirel had suggested a couple of weeks before the military coup d'état that the question of command and control rights might be solved after Greece had returned to NATO. However, this proposal had faced severe criticism from the leftist and Islamic parties.

After the reintegration had taken place, Turkish officials declared that this was in the interest of Turkey and welcomed Greece's return (cf. NYT, October 21st, 1980). On the other hand, Greek government sources were reported to have stated that the Turkish junta gained nothing concrete from the agreement and only gave its consent because of the deteriorating international situation and the need to concentrate on domestic matters.

[106] Güldemir asserts that after the military takeover Turkey and the US were as close to each other as ever since the conclusion of the Baghdad Pact (cf. Güldemir 1986, 115-117). Güldemir lists a number of developments confirming that instead of causing a break in US-Turkish relations, the military takeover induced a period of close cooperation. For example, Turkey reportedly agreed to the continuation of U-2 flights over the Soviet Union, which were denied during the Premiership of Demirel. On September 29th, 1980, the International Monetary Fund (IMF) released $ 92 million in aid to Turkey. On October 25th, 1980, the US agreed to postpone the repayment of a $ 350 million loan for a year. A consortium of 16 banks followed the US practice and decided to postpone the repayment of loans amounting to $ 3.2 billion for a period of 7 – 10 years. In spring 1981, the Pentagon declared it had decided to sell 15 F-4E fighter jets worth $ 78 million to Turkey. Above all, in April 1982, Turkey and the US reached agreement on measures to be taken in order to enhance Turkish defense capabilities. The agreement envisaged, among other measures adopted for the modernization of the Turkish armed forces, the production of 160 F-16 fighter jets in Turkey from 1987 onwards and the modernization of 16 airfields.

[107] In an interview with Güldemir, the US President of the time, Jimmy Carter, stated that the problems related to Greece's return to NATO were solved thanks to the friendship between Rogers and Evren. He added that this problem could not have been solved if the coup of September 12th had not occurred (Güldemir 1986, 94).

Between September 12th and October 17th when Turkey agreed to lift its veto against Greek reintegration, General Rogers visited Turkey a total of four times. Turkey finally accepted his proposal to allow Greece's reintegration first and to resolve the issues related to the command and control rights thereafter (Ibid., 78). The military regime declared that the decision had been taken to facilitate the refilling of the gap that had emerged in the defense line between Turkey and Italy after Greece's withdrawal from NATO (Ibid., 80). They insisted the lifting of the veto reflected Turkish interests and should not be regarded as a concession. However, the Rogers Plan had no binding character and the only assurance Turkey received was a promise by General Rogers. Thus, when Papandreou refused to implement the part of the Rogers Plan concerning Greece and tried to use NATO forums to gather support against Turkey, there was nothing Turkey could do.[108]

Leaving aside the question of what had moved the military regime to accede to Greece's reintegration, it is a fact that the veto was lifted only five weeks after the military had seized power without any major concessions from Greece, and there were no guarantees except for a promise given by the SACEUR for safeguarding Turkish interests. Thus the decision to lift the veto constituted a major and abrupt deviation from the policy Turkey had been pursuing since 1974.

---

[108] According to Güldemir, even though not included in the blueprint of the agreement, there was an unofficial understanding that Turkey would be granted certain side-payments due to its lifting of the veto (1986, 83-84). A lifting of the veto would contribute to the standing of the regime in Washington, and Turkey's defense requirements would be paid greater attention. Rogers would personally work towards this end. Furthermore, as the agreement did not envisage any measures to be adopted in case of a Greek violation of the agreement, Rogers assured Turkey that he and the US would intervene on behalf of Turkey in such a case; that is to say, the only guarantee Turkey received was a "soldier's promise."

During a meeting with journalists on October 27th, 1985, Kenan Evren claimed it was not true that he had told Generals Rogers they would accept Greece's reintegration if they were in power (cf. Güldemir 1986, 135-139). There was not a linkage between the prospect of additional US aid and the lifting of the veto against Greece. According to Evren, Turkey had proposed the formula according to which Greece and Turkey were to declare they had no command and control responsibilities in the Aegean and Greece was to return to NATO thereupon allowing for negotiations on the distribution of rights already in March 1979. However, this proposal had been rejected both by General Haig and Greece. Thus the Rogers formula, in the view of Evren, reflected the Turkish proposal and Greece's readmission subsequently occurred on terms favorable to Turkey. Evren added that no civilian government in Turkey was opposed to Greece's reintegration and that at the point of time when the military seized power the issue had almost been solved. Hence, he offered the view that Greece would have been admitted to NATO even if there had not been a military coup in Turkey.

One should also add that Güldemir cites interviews with General Rogers, General Haig, US Ambassador James Spain, and President Jimmy Carter as sources upon which his *thesis* rests.

Even so, the lifting of the veto did not constitute the only deviation from previous Turkish policies towards Greece. Özcan adds that policies towards Greece were now formulated in a manner that would help to avoid any disturbances within the Atlantic Alliance (cf. Özcan 2001a, 517). The desire of the Turkish generals to maintain good relations went so far that press articles critical of Greece were censored. In 1982, the General Staff told the representatives of the Turkish press that relations with Greece had to be dealt with in a less prominent manner. After the coup, the Americans had also asked the military regime to show greater flexibility in Cyprus in order to help the inter-communal talks, in progress again since 1980, to bear fruits (Stearns 1992, 121). After some delay, Turkey was to submit proposals for reducing the territory under Turkish control from 36 to 30 percent in August 1981. As new elections were scheduled in Greece for October 1981, the Rallis government refrained from responding to the Turkish offer.

As military regimes in Turkey were usually inclined to act in a more conciliatory manner towards the neighbor to the west,[109] the Colonels of Greece, too, pursued a rapprochement with Turkey in similar attempts to improve their image in Washington and in European capitals. Coufoudakis asserts that the Greek junta leader Papadopoulos had been a proponent of friendship with Turkey and believed it was possible that both countries might unite within the framework of a federation one day (1985, 197). Given this desire of the junta to improve ties with Turkey, the Greek and Turkish prime ministers held a secret meeting at the Greek-Turkish border five months after the coup d'état had taken place in Greece (cf. Ierodiakonou 1971, 269). The Colonels had been hoping that they could achieve enosis by providing side-payments to Turkey (cf. Uslu 2000, 223-224). If accepted by Turkey, they would be able to fulfill a centuries-old dream and put an end to the conflict with Turkey, simultaneously earning the appreciation of the Americans. Hence at their meeting, they offered modifications of the Greek-Turkish border in exchange for Turkey's consent to enosis (Uslu 2000, 204). The Turks, however, rejected the offer as the side-payments offered by the Greeks were regarded as insufficient. Still, even though they had failed to agree on a solution for Cyprus, they committed themselves to improving their relations and decided to continue their efforts to find a solution to the Cyprus question. However, the two countries were to find themselves at the brink of war two months later when the National Guard attacked the Turkish Cypriot village of Kophinou leaving 24 Turks dead and one Greek wounded (Ierodiakonou 1971, 269).

---

[109] Paradoxically, the Turkish military is known for its hard-line policy vis-à-vis Greece (cf. Stivachtis 2001, 40).

Nonetheless, the efforts at improving ties were to continue during the reign of the Colonels. Stearns, for example, attributes the 1971 agreement between Greece and Turkey to promote the resumption of inter-communal talks and to impose a definitive solution if the communities in Cyprus failed to work out a formula on their own to the facts that the new government in Turkey installed by the military adopted a more facilitative role while the Colonels' regime was interested in improving its standing in the Western world and viewed gestures of good-will towards Turkey as the best way to achieve this end (Stearns 1992, 114-115). According to Stearns, substantial progress had been made in the period of 1971-1972, and the parties had almost reached agreement. However, due to the divisions within the Colonels' regime and between Athens and Nicosia[110] on the policy to be pursued towards Turkey, Greece's support for the talks had been dwindling and General Grivas' return to Cyprus to continue the struggle for enosis, which forced Makarios to adopt a less conciliatory role, burdened the talks which then came to a halt in the face of the developments of summer 1974 (Ibid., 116).

Returning to relations between Turkey and the US, in the post-Cold War period the latter seems to pay more attention to the human rights situation and to the state of Turkish democracy (Uzgel 2001c, 252). For example, in 1994, Congress declared that aid to Turkey must not be used in violation of international law and made 10 percent of the aid contingent on the report of the State Department on the progress made in Turkey with regard to human rights and the Cyprus issue. The report of the State Department said Turkey violated human rights in the campaign against the PKK but added that Turkey had the right to defend itself in order to guarantee its internal security. Turkey, nevertheless, rejected the part of the aid made contingent on the report (cf. Uzgel 2001c, 286). In a similar manner, the number of Cobra helicopters to be sold to Turkey was reduced from 50 to 10 in the face of opposition by human rights groups and Congress in 1995 (Ibid., 288).

---

[110] According to Uslu, the junta and the Turks were close to a solution on the basis of double-enosis in 1971 (Uslu 2000, 225). Makarios, on the other hand, did not want enosis as long as such concessions had to be made to the Turks (Ibid., 227). He insisted on enosis and rejected the idea of double-enosis which would have meant giving up his claims to one part of Cyprus. Thus, the independence of Cyprus should be preserved up to the point where Cyprus as a whole could join Greece. In the view of the junta, however, Makarios had betrayed the cause of enosis and was hostile to the regime in Athens as such. Ioannides, who seized power after a coup against Papadopoulos in 1973, was ever more committed to the goal of enosis (Uslu 2000, 230). According to Birand, it was Ioannides who controlled Sampson and the coup against Makarios was carried out without consulting other members of the junta, which caused discomfort among these junta members (1976, 53, 55).
It is also worth mentioning that Greek-Turkish relations had been burdened by the tensions of 1973 related to the shelf issue (Coufoudakis 1985, 197).

The same considerations had led to delays in the delivery of frigates[111] and tanker-planes leased or rented by Turkey. Uzgel maintains that Turkey, nonetheless, continued to receive large amounts of US arms despite these difficulties and delays. The US was interested in preserving its ties with Turkey - ties that constituted an important factor in US policies towards the Caucasus, Central Asia, and the Middle East. Last but not least, Washington was apparently interested in maintaining a big share in the armament program of the country (Ibid., 288).

Ironically, Washington sometimes seems to miss the periods of harmony in relations between the two countries achieved during the tenure of military regimes. The remarks of a high-ranking US official during the war in Iraq perfectly illustrate this point. The attack was to cause friction between Turkey and the US since the Turkish parliament did not give its consent to the deployment of US troops in the country, an act that would have enabled the latter to open a northern front against Iraq. In May 2003, the Deputy Secretary of Defense, Paul Wolfowitz, said that the Turkish military had not played "the strong leadership role on that issue we would have expected" (cf. TDN, May 23rd, 2003) during an interview on TV. Wolfowitz's remark caused an uproar as it was interpreted as an expression of regret that the military had not forced the parliament to allow the deployment of US troops in Turkey. Some of this criticism was also rooted in the US. The Republican Senator Barney Franks called on Wolfowitz to resign since his statement could undermine democracy in Turkey. Franks added it was "appalling to have such a high ranking American official say this" (TDN, May 23rd, 2003).[112]

---

[111] In 1997 Congress blocked the delivery of three frigates to Turkey on the grounds of human rights violations in that country (TDN, July 25th, 1998). However, these frigates were delivered in 1998.

[112] It was also interesting to see the US force Belgium to amend a law introduced in 1993 which empowered Belgian courts to try suspects of genocide or other war crimes irrespective of their nationality or the location where the crime was committed (cf. IHT, June 24th, 2003). After the US/UK invasion of Iraq had begun, cases were filed against US President Bush, Secretary of State Powell, Secretary of Defense Rumsfeld, UK Prime Minister Blair, and General Tommy Franks, the commander of the forces in Iraq. The US warned that the lawsuits might force the US officials to boycott the meetings of the NAC in Brussels. Rumsfeld was even more outspoken and said that NATO might have to move out of Brussels. Meanwhile, he froze funding for the new headquarters to be built in Brussels. Belgium had to bow to US pressure and amend the law in question. The new legislation restricted the jurisdiction of the law to those countries that lacked a functioning and independent judicial system (NZZ, June 23rd, 2003). Additionally, only those cases directly linked to Belgium would be dealt with in Belgian courts in the future. Apart from this bilateral dispute with Belgium, the US froze military aid to a number of states that favored the creation of the International Criminal Court and refused to accept special treatment for US citizens (NZZ, July 2nd, 2003). US President Bush also warned the Europeans that the US role in NATO would change unless the EU agreed to the US proposal to exempt Americans from the jurisdiction of the Court

With regard to the state of democracy in Turkey today, Semih D. İdiz writing for the *Turkish Daily News*, held in 1997 that the country continued to constitute an "oddity" in NATO (TDN, September 17th, 1997). It is especially the dominant position of the military, which perceives itself as the guarantor of the secular system, that gives rise to criticism at home and abroad. In fact, the former Chief of General Staff, Doğan Güreş, is said to have stated in 1992 that Turkey was a military state (Özcan 2001b, 16).

The main body allowing the generals to exert influence on the domestic and external policies of the country is the National Security Council (Milli Güvenlik Kurulu, MGK). It is the MGK that produces the "National Security Policy Document" which "lists the threats to national security, sets out priorities, lays down policy guidelines, and provides a detailed framework of foreign and security policies for governments and state institutions" (Özcan 2001b, 20). All institutions of the state are required to carry out their duties in conformity with the provisions of this paper (Ibid., 21). As a consequence, an unaccountable body dominated by the military works out the parameters of policy-making and implementation in Turkey. This is why this document is referred to as the "secret constitution" in Turkey. Moreover, despite the fact that the council is *de jure* an advisory body and the constitution explicitly states that the decisions taken by the council are not binding, *de facto*, there is no doubt among civilians and the military that its decisions are binding (Kramer 2000, 34). The council can formulate "recommendations" on any security issue. As the concept of security is defined in broad terms, almost any matter can be put on the agenda of the council, which is mainly shaped by the generals (Özcan 2001a, 524-525). Correspondingly, it is worth mentioning that the Turkish military is especially sensitive towards issues touching on the secular character[113] of the system and the territorial integrity of the country. Thus, with regard to religious freedom and the fate of the Kurdish citizens, no policy changes can be undertaken without prior approval by the military.

The military's role is also not confined to internal matters. It plays an important and sometimes independent role in the formulation of foreign policy (Özcan

---

(IHT, August 27th, 2002). Finally, in July 2002, the UN reached agreement that US peacekeepers should be exempted for a one-year period.

[113] The military sometimes appears to be *oversensitive*. For example, military officers and the president boycotted a reception hosted by Parliament Speaker Bülent Arınc because his wife, who wears a headscarf, was expected to participate in the reception. Moreover, a number of officers are expelled from the TSK each year due to "undisciplined behavior." a term which in fact refers to Islamic reactionary activities as the *Turkish Daily News* puts it (cf. TDN, August 2nd, 2003). According to a report by the TDN, "most of the time, such activities refer to the headscarf of the wives of the personnel in question."

2001a, 533)[114] - indeed this is sometimes true for the presidency and the foreign ministry as well (Özcan 2001b, 14-15). The most striking example of this independent role played by the military is the strategic partnership between Israel and Turkey. This is a project brought to life by the military, which reached its peak in 1996 with the conclusion of two accords on military training and cooperation in the defense industry despite the opposition of the Erbakan government (Özcan 2001b, 19, 23). The signing of these accords was in sharp contrast to the civilian government's efforts to forge closer ties with Muslim countries like Iran or Libya. On another occasion in May 1997, the government was informed about the operation of Turkish troops beyond the border in Iraq only after the troops had already crossed the border (Özcan 2001b, 23).

The military is also known for its intransigent stance towards Greece and is highly sensitive on the issue of Cyprus. Writing in 2000, Bahçeli maintained that the hard-line stance of the military on Greek-Turkish relations "discourage[d] a change of policy towards Athens" (Bahçeli 2000, 466). She added, however, that other institutions like the foreign ministry shared the views of the military while both were aware of the potential costs involved in a war with Greece. Chipman, in a similar fashion, notes that the politicians may negotiate a settlement, but on military aspects "they will take their cue from the armed forces" (1988d, 369). To cite an example, the general staff was reported to have sent a letter to the foreign ministry urging the latter to change the "passive" policies related to the Aegean Sea in 1997 (Özcan 2001b, 22). The military had also declared its opposition against the first versions of the Cyprus plan put forward by Kofi Annan in November 2002 (NZZ, January 10th, 2003). The Chief of General Staff, Hilmi Özkök, described the plan as a threat to the security of Turkey and added that no plan disregarding the security interests of Turkey could guarantee durable peace. While the government seemed to be willing to accept a solution on the basis of the Annan Plan, and Premier Erdoğan criticized Turkish Cypriot leader Denktaş and his Greek Cypriot counterpart Clerides for lacking the will to solve the issue (TDN, January 28th, 2003), the commander of the Turkish Land Forces, General Yalman, visited Denktaş in a show of support and argued that the plan put forward by Annan could lead to violence similar to the circumstances of the 1960s. Still, the military seemed to have revised its stance by January 2004 when

---

[114] The National Security Policy Document referred to Greece as one of the most prominent threats to Turkey's security for the last time in its 1997 revision. This document, which was brought up to date in 2001 during the tenure of Bülent Ecevit, is presently undergoing further revision (cf. Hürriyet, November 25th, 2004; Radikal, November 28th, 2004). Given the atmosphere of recent years, Greece is not expected to be put on the list of major threats to the country if listed as a threat at all.

the MGK issued a declaration stating that a solution had to take the Annan Plan as a reference point and be based on the realities of the island. The *Turkish Daily News* wrote that "[i]t could be said that after 14 months in power, the AKP [PM Erdoğan's Justice and Development Party] has finally become the real government of Turkey because with the wording of the statement it has demonstrated its 'ability' to promote its position - albeit with some compromise - with the MGK" (TDN, January 26th, 2004). According to the report, "[a]t Davos the Turkish prime minister demonstrated his ability of pushing for a 'compromise' settlement on a very sensitive foreign policy issue despite the 'expressed' opposition of some top commanders and some important sectors of Turkish society as well as of Turkish Cypriot President Rauf Denktaş, who has more popularity in Anatolia than any Turkish politician." After the talks in Switzerland had ended, the MGK endorsed the final version of the plan and declared the government responsible for safeguarding Turkish interests with regard to the formalization thereof (Hürriyet, April 5th, 2004).

For the sake of fairness, one has to add that there have also been instances where the military tried to play the leading role in improving ties with the neighbor. In 1993 the Chief of General Staff, Güreş, sent a letter to his Greek counterpart suggesting both militaries should make the first steps towards peace between the two countries (Özcan 2001a, 519). In addition, Güreş proposed a summit meeting of the chiefs of general staff of the Balkan countries. Apart from that, in 1999, the commander of Turkish military academies, General Necati Özgen, declared that the Aegean disputes required a quick resolution while the provisions of international law should also be taken into account (Riemer 1999, 555). The former Chief of the Turkish General Staff, General Kıvrıkoğlu, announced in April 2000 (TDN, April 25th, 2000) that they wanted to start a dialogue with the Greek military. He added they had made a number of proposals such as holding joint maneuvers or reducing the number of exercises, but the Greeks had still not responded to these proposals. The most recent gesture of friendship was made by the current Chief of Staff, Hilmi Özkök. Özkök, holding a speech during the SEESIM 2004 maneuvers which Greek officials also attended, maintained that Turkey no longer perceived any country as a threat to its security (cf. CNN Türk, November 12th, 2004). He added that the major threat all countries had to deal with was terrorism. Evaluating the statement made by the Chief of Staff, an article on the website of the Austrian Broadcasting Company, ORF, stated, "Turkey ends enmity with Greece" (ORF.at, November 12th, 2004).

Returning to the question of the military's role in foreign policy-making in general, Özcan argues that the military's influence increased in the 1990s (2001b, 13, 25). This process was facilitated by geopolitical changes, the flaring up of

Kurdish separatism, and what he calls Islamic fundamentalism. In this period, the military concluded agreements with 30 countries in the region surrounding Turkey, thus further enhancing its role in the foreign relations of the country. However, Özcan adds that non-governmental organizations (e.g. Association of Turkish Industrialists and Businessmen) also played a greater role, and offers the view that the military's visible role in charting foreign policy might diminish once relations with neighboring countries like Syria, Iran, and Greece have improved, the PKK has been brought under control, and the prospects of becoming a EU member-state have been enhanced (Ibid., 26). Today, most of these conditions seem to be present: the PKK has been defeated, relations with Greece have improved, and a cautious rapprochement process seems to be in progress with Syria. Indeed, the current Chief of General Staff, Hilmi Özkök, who was the military representative of Turkey at NATO for many years, is said to be in favor of democratic reforms and willing to curb the military's role in civilian affairs (WP, April 9th, 2003). Özkök stated during an interview that the armed forces supported the goal of EU membership and the reform process to this end. The top commander argued that 70 percent of the Turkish population supported membership and that in the face of such support nobody could oppose it. He also added that the problems with Greece could be solved within a week if Turkey were to become an EU member (TDN, October 20th, 2003).[115] Nonetheless, he is reportedly not interested in abandoning the military's traditional stance against the issues related to Kurdish separatism and Islam.

In recent years, a reform process has been in progress in Turkey, mainly in an attempt to fulfill the accession criteria put forward by the EU. A total of 34 amendments were made to the Turkish constitution in October 2001 (cf. Axt 2002, 48). Among others, the death penalty was abolished, except for terrorists convicted of treason, although the death penalty was later abolished as a whole

---

[115] A certain degree of skepticism exists about whether the military has really embraced the idea of EU membership. Riemer, for example, claims that the army is not interested in membership, since this would drastically constrain its ability to exert influence on the political system (cf. Riemer 1999, 549). According to Riemer, the army equates western orientation with an orientation towards the US - whose main objective is the preservation of stability in Turkey irrespective of the means and strategies involved to achieve and maintain this end (Riemer 1999, 557). Thus there is no need and desire to join the EU and give up its dominant position in the countries' affairs (see also Riemer & Stivachtis 2000, 565).

In fact, according to an article printed in the *Washington Post* (April 9th, 2003), there are certain factions in the Turkish Armed Forces favoring different foreign policy orientations. A group formed around the former Chief of Staff Kıvrıkoğlu and the commander of the land forces General Yalman is said to prefer establishing closer ties with nations such as Russia and China. These officers are reported to be suspicious of Europe and the US, in great part because of those countries' support for the Iraqi Kurds. They fear that the US backs the idea of a Kurdish state which might serve as a "reliable source of oil".

(TDN, January 8th, 2004); some restrictions concerning parties and trade unions were lifted; and the number of military members of the National Security Council (MGK) was reduced to 5 while the number of civilians was increased from 5 to 9. Moreover, an explicit reference was made in the constitution to the non-binding character of MGK decisions. In addition, laws from the time of the military regime could now be taken to the courts. In July 2002, the emergency state was lifted in two provinces in Southeast Anatolia (Axt 2003, 68). In the same year, among other allowances, television broadcasts in the Kurdish language were permitted and private institutions were allowed to offer Kurdish language courses. Freedom of speech was extended, so that criticizing the military or any apparatus of the state could no longer be punished. Apart from that, organizations of minorities and religious communities were now able to acquire and sell properties.

In 2003, the government passed the seventh harmonization package further curbing the influence of the military (cf. TDN, August 2nd, 2003). The role of the MGK as an advisory body was emphasized, the frequency of council meetings was reduced from meeting once every month to bimonthly meetings, and the parliament was given more room for scrutinizing military expenditures that had previously attained automatic approval by the parliament. Moreover, a civilian, nominated by the prime minister and appointed by the president, is now allowed to head the secretariat of the council, , although the chief of staff is to be consulted when the nominee is a military official (see also Hürriyet, July 31st, 2003).[116] In addition, the task of overseeing the implementation of council decisions adopted by the council of ministers was transferred to the deputy premier. Furthermore, civilians can no longer be tried before military courts for offences such as discouraging people from carrying out their military service in peacetime. The MGK was also deprived of the right to determine which languages could be taught (Radikal, July 31st, 2003). What is more, allegations of torture must now be dealt with without delay, and some provisions restricting the freedom of speech and assembly were lifted (IHT, July 31st, 2003; August 4th, 2003). The *Financial Times* referred to the reforms as a "quiet revolution" if implemented (quoted in the TDN, August 2nd, 2003).

The reform process has not come to a standstill in the following period. The military members were removed from the Higher Education Board and the Supreme Board for Radio and Television, and the State Security Courts were abolished (TDN, May 24th, 2004). In September 2004, the Turkish Parliament passed the new penal code of the country in an attempt to comply with EU norms (Hürriyet, September 29th, 2004). EU's outgoing commissioner for enlargement,

---

[116] Meanwhile, former Turkish Ambassador to Greece, Yiğit Alpoğan, became the first ever civilian secretary of the MGK.

Günther Verheugen, referred to the new penal code as a "Jahrhundertwerk"- "a work for the century"- which was inevitable for the beginning of accession talks (cf. NZZ, September 26th, 2004). The new penal code is expected to enforce the rule of the law and freedom of speech. A new law on intermediate courts of appeal was also adopted. The Turkish parliament also recently passed a new code on criminal procedure. Among other changes, the new regulations cut the powers of the police while extending those of suspects or detainees (cf. Kathimerini, Radikal, TDN, December 6th, 2004). For example, suspects can no longer be detained for more than 24 hours without charge and they have to be informed on their rights when arrested. Moreover, the police will have to ask the judiciary for permission in order to be able to conduct house searches. Having passed the aforementioned laws, the parliament is expected to start elaborating on the revision of the law on execution of punishments soon.

Notwithstanding the fact that the country has taken serious steps towards fulfilling certain democratic standards, which would not have been possible without the military giving its consent or, at least, condoning this process, the question remains as to how far these reforms will go and whether they will be fully implemented. The striking question will be whether the military will concede to the principle of civilian supremacy not only on paper but in reality, too. The former President of the EU Commission, Romano Prodi, made this point when he stated that not only the direction of the reforms was correct but the speed was remarkable as well. He added though that the implementation of these reforms was essential to meet the Copenhagen Criteria (TDN, January 17th, 2004).

About 9 months after Prodi's visit, the outgoing EU Commission certified Turkey it had fulfilled the political criteria and recommended the opening of accession talks.[117] With regard to the role played by the military, the Commission offered the view that "civil-military relations [were] evolving towards European standards" (2004c, 3), but "the armed forces in Turkey continue[d] to exercise influence through a series of informal mechanisms" (2004c, 11). Nonetheless, what Turkey received was, in the words of Prodi, "a qualified yes" (IHT, October 7th, 2004). While praising the reform process in progress and pointing to the accomplishments of recent years, the Commission also drew attention to the fact that "implementation need[ed] to be further consolidated and broadened." Or as put by Günther Verheugen, "[t]he legal framework does correspond to a democratic state of law but practice does not" (IHT, October 7th, 2004). Yet, the Commissioner also added that no accession talks should have been opened with

---

[117] For the "Recommendation of the European Commission on Turkey's Progress towards Accession" see the EU website: http://europa.eu.int/comm/enlargement/report_2004/pdf/tr_recommendation_en.pdf

any of the ten countries which acceded to the EU in May 2004 had the Commission insisted on a prior one hundred percent implementation of reforms (ORF.at, October 6th, 2004). Meanwhile, the reform process is not regarded as being "irreversible," inducing the recommendation of certain safeguards by the Commission. In line with this recommendation, the negotiations with Turkey can be suspended should the country backtrack on democratic reform or violate fundamental principles the Union is founded upon. On the whole, despite the necessity for further reform and quick implementation, the reform process in Turkey deserves credit and the recommendation provided by the Commission confirms the progress made by the country.

However, the fact that Turkey has undertaken serious steps towards fulfilling certain democratic standards in recent years does not change the reality that almost five decades of membership in NATO have failed to move the country towards further democratization. Indeed, it is obvious that the main impetus for the recent reform process came from the EU, even though internal pressures for democratization should not be overlooked. While the desire to obtain US aid had facilitated moves towards a multiparty system, the conditionality of US help fell short of giving rise to a truly democratic system in the country, and membership in NATO did not prevent the military from carrying out 3 ½ coups. In fact, it turned out that the Alliance's leading force, the US, and the Alliance itself had no inclination to put their relations with the country at risk, enjoyed close relations with the military regimes, and, at times, found it easier to cooperate with the military than with civilian governments. To put it in other words, the country usually did not face serious pressures to democratize and had no reason to fear abandonment when the military periodically intervened. It is obvious that no socialization process or an indoctrination of democratic values have taken place in spite of close contacts between the Turkish military officers and their Western, especially US, counterparts. Despite recent developments, Turkey remains a country lacking full civilian control of the armed forces.

On the other side of the Aegean, there is no real indication that Greece does not fulfill Western standards related to the civilian control of the armed forces today. Greek democracy is generally viewed as consolidated - notwithstanding the fact that Greece has, at times, had to face criticism because of its treatment of the Turkish minority. However, the process of consolidating Greek democracy was initiated in the post-junta period, that is to say, after Greece had left NATO's military structure. Neither the prospect of membership, nor the education and training of a great number of Greek officers by NATO and US officers had convinced these officers to accept civilian supremacy. Using NATO plans, they had seized power in 1967 and ruled the country for seven years. In order to gain the favor of the Americans, they had also been more than receptive towards US wishes. Strategic necessities had once more determined the relations of NATO and the US with the junta in spite of the fact that certain NATO members had wanted to distance themselves from the regime. However, given the fact that their authoritarian regime had alienated the public, brought much pain on their compatriots in Cyprus, and had almost drawn the country into a catastrophic war with Turkey,[118] the Greek military was no longer in a position to reassert itself in the post-1974 period and had to adhere to the principles of civilian control and the primacy of democratic rule.

On the whole, with regard to the democratic development of Greece and Turkey, NATO membership does not seem to have made any major difference. While both countries viewed the Alliance as a democratic club and thought membership would help to consolidate their fragile democracies, the prospect of membership did not give rise to the emergence of real democracies, nor was there any socialization process making the militaries of these countries accept civilian supremacy. Meanwhile the Alliance and the US in particular turned out to be perfectly able and willing to work with military regimes in these countries. Ironically, cooperation with military regimes, at times, even proved to be easier than with elected governments.

---

[118] In 1975, the US Vice-President Nelson Rockefeller stated that Greece should be grateful to Turkey for overthrowing the junta (Woodhouse 1982, 254). Nonetheless, President Clinton publicly apologized for US support to the junta (Clogg 2002, 231).

6) CONCLUSION

The aim of this book has been to shed light on the effects of Greek-Turkish membership in NATO on their bilateral conflict as well as on their democratic development while paying attention to the ongoing debate between representatives of different schools of international relations theory and focusing the analysis on such *institutional factors* like issue linkage, transparency, or consultations. Thus, having outlined the course of relations between Greece and Turkey and depicted the issues of conflict, the effort has been undertaken to portray the assessments of (neo)realists and neoliberal institutionalists. The discussion of these theories has shown that they have to be viewed as complementary rather than as contradictory. Both theories pay attention to concerns related to relative gains or cheating, but neorealists focus on situations where the incentives for overcoming such concerns are weak, and tend to view the world in more competitive terms. One of the consequences of this focus on situations less favorable for cooperation has been that neorealists have neglected developing a theory of institutions, as acknowledged by Grieco. On the other hand, institutionalists admit that cooperation cannot occur under any circumstances. Moreover, the effects of institutions will vary in different settings. They are embedded in the realities of power and interest and should not be expected to have the same positive effects under all conditions. Hence, institutional *factors* like transparency or issue-linkage might facilitate cooperation just as they can impede the occurrence of such cooperation. On the whole, the validity of neorealist or institutionalist assumptions is dependent on the situational setting at hand while the same principle applies to the effectiveness of institutions as facilitators of cooperation and stability.

Returning to the effects of more specific *functions* of institutions such as transparency, consultations, issue-linkage, or to the phenomenon of institutional capture and the consequences arising out of military assistance provided to the feuding members, the findings of this book hint that NATO has sometimes helped to mitigate the conflict, while on other occasions, factors related to the Alliance have contributed to existing tensions.

To begin with, Krebs' argument, according to which the Alliance has added to the issues of discontent and both parties have captured the Alliance to further their own positions vis-à-vis the opponent, seems to be accurate - it is worth remembering that this does not contradict institutionalist assumptions. Disagreements related to command and control rights in the Aegean, NATO infrastructure funds, the inclusion of certain Greek islands in NATO maneuvers, and the assignment of certain troops to NATO have emerged as new items on the list of Greek-Turkish problems. Furthermore, especially Greece, the weaker of the two sides, has made use of NATO forums to gather support against Turkey and established a linkage between the extent of Greek cooperation and the attitude adopted by the Alliance in the face of their bilateral problems. Meanwhile, linkages drawn between issues of dispute and other issues have usually proved destructive. Yet, policy-makers in both countries as well as NATO officials seem to be aware of this fact. NATO officials maintained that the Alliance tried to isolate or find ways to work around contentious issues to mitigate their repercussions on the functioning of the institution (Interviews in Brussels, July 2001). The objective of the Alliance seems to have been to prevent *destructive linkages* and establish *islands of cooperation,* to use the terms of Hasenclever, rather than trying to solve the contentious issues by providing side-payments through issue-linkages. Greek and Turkish officials have also adopted the strategy of unlinking contentious issues from items where agreement seemed possible when they wanted to facilitate cooperation given the difficulties they have had in sorting out their problems for decades. They have also been following this path ever since a new period of détente began in the summer of 1999. Leaving aside their bilateral problems, they have tried to further cooperation on issues like tourism, illegal immigration, and trade, hoping this would contribute to a better atmosphere between the two countries allowing them to tackle the contentious issues later on.

To arrive at determinate conclusions on the effects of military aid seems to be more difficult. Military assistance provided by the Alliance, and the US in particular, is said to have created an imbalance favoring Turkey and simultaneously exacerbated the internal security-dilemma. According to Krebs, when two countries are in conflict, military aid adds to the insecurity felt by the parties and feeds mutual suspicions. As a consequence, adversaries refrain from making concession in an attempt to display resolve anticipating revisionist aims behind the weapons acquisitions. However, while such acquisitions might have exacerbated mutual suspicions in the case at hand, they might have added to the deterrence capabilities of both parties as well, preventing them from going to war against each other in times of crisis. Moreover, both states would probably have sought to acquire weapons even if they had not been members to the Alliance. It is open to speculation what their arsenals and force postures would have looked like if these countries had remained outside the Alliance. What is beyond doubt is that they would have lacked the communication channels to clarify misperceptions when necessary. Apart from that, it is hard to conceive of how a balance could be established between two countries that are unequal in size, population, and internal and external threat perceptions. On the whole, while military assistance by the US or NATO might have added to tensions, it is hard to imagine that they would have escaped this fate had they not been members of the Alliance. Meanwhile, such military acquisitions might simultaneously have facilitated stability contributing to mutual deterrence. In fact, there seems to be much room for speculation on this issue.

The effects of transparency seem to be ambiguous, too. NATO provides for a certain degree of transparency on capabilities. However, transparency on capabilities need not necessarily alleviate concerns of other members and might indeed work to the other end. Yet, when talking about transparency institutionalists refer to capabilities as well as intentions and expectations. Consultations within the framework of institutions are expected to provide allies with an insight into the intentions of others and enable member-states to assure and reassure each other. In cases where misperceptions are the problem, such transparency can be expected to be of help. However, when military *intimidation* or even revisionist aims have conditioned the acquisition of certain capabilities or force postures, transparency cannot be expected to facilitate cooperation and might even have adverse effects. Thus, while sharing information on the capabilities of the other member or having channels of communication to clarify misperceptions or to gain insight into the intentions of the other party must still be valuable assets for Greece and Turkey, transparency *per se* should not be viewed as a panacea against conflict, its effects being dependent on the nature of the relationship at hand.

With regard to the effects of NATO membership on the democratic development of Greece and Turkey, a belief in the Alliance's character as a democratic club and the expectation that membership would help to consolidate their fragile democracies seems to have contributed to their desire to join NATO. However, in retrospective, NATO membership seems to have played no major role in shaping the fate of democratization in these countries. On the contrary, strategic considerations have usually outweighed the allies' concerns for democratization, and NATO and the US have been able to work with military regimes in these countries perfectly well (see also Tovias 1991, 176-177; Segal 1991, 40). It is also obvious that there has been no socialization process, which would have resulted in an internalization of democratic principles by the military officials from these countries. Despite the fact that many military officers from Greece and Turkey had been educated at the NATO College or in the US, they had no hesitations when it came to seizing power or interfering in politics. Given this failure of socialization processes and the fact that there are no treaty provisions and no institutional mechanisms that would allow the Alliance - which works by consensus - to enforce democratization or punish deviations from democratic principles, NATO membership should not be viewed as an effective means for facilitating democratization in member-states. Indeed, NATO officials seem to be aware of this fact, and the strategy adopted by NATO has been to move aspirant countries towards democratization by making membership contingent on their democratic record (Interviews in Brussels, July 2001). As one NATO official put it, once countries have joined NATO, it is impossible to effect change. Concerning the case of Turkey, he added that if the Alliance had pressed for democratization in Turkey, it would have probably distanced itself from the Alliance rendering it even more difficult for the Alliance to exert influence on the country. According to the official, there has been a division of labor between the EU and NATO. While the first has been pressing for democratization taking into account that this might harm the links with the country in question, the latter has chosen to refrain from criticizing it for its democratic record and left the job to the EU. However, one should add that this division of labor has also been driven by self-interest. Given the EU's reluctance to extend membership to Turkey, the country's bid for membership could be turned down by pointing to democratic deficits. Simultaneously, as Turkey has been valued as a strategic partner, the European countries have had to refrain from alienating the same country in NATO. Overall, while it is plausible to think that the prospect of membership might facilitate democratization in aspirant countries, the cases of Greece and Turkey show that once countries have joined the Alliance, they will have no major inclination to beef up their democratic record.

On the whole, regarding the main purpose of this thesis, which has been to find out what the effect of membership in NATO has been on the course of relations between Greece and Turkey, the findings hint that, overall, factors unrelated to their membership in NATO have determined the development of this conflict. In 1974, Greece would probably have gone to war against its NATO partner had it been in a position to do so. Turkey, meanwhile, had been expecting such a move by Greece and took measures to counter an attack. Thus, the view offered by Kurop (1998, 11) according to which NATO membership is understood as an implicit non-aggression guarantee is not accurate. Simultaneously, efforts to improve ties have usually been conditioned by factors unrelated to NATO, the recent rapprochement process constituting a case in point. The point to be made is that NATO's role in this conflict has never been a prominent one. Nevertheless, membership in NATO has had certain effects on the course of Greek-Turkish relations, yet these effects have been ambivalent in nature. As argued by Krebs, NATO membership has, on the one hand, led to an extension of their conflict into new areas and, at times caused additional tensions. On the other hand, NATO has served as a face-saving forum where officials from both countries could meet, discuss problems, or clarify misperceptions. At times, the institution has served as the only channel for communication between the two countries. Moreover, NATO secretary-generals have successively offered their offices to the parties and tried to defuse tensions, even though the net effect of such mediation efforts is questionable. NATO membership can also be seen as a factor increasing the political and diplomatic costs of waging war against an ally, though not rendering it impossible as already stated.

Indeed, NATO does not seem well suited for dealing with internal conflicts. As a NATO official put it, it is an institution working by consensus and can thus not be expected to impose any solution or adopt any measures against a member-state. NATO cannot dictate anything, as another official put it, because these states themselves constitute NATO (Interviews in Brussels, July 2001). Another official cited the case of an official working for NATO who had been criticized for a remark made during the enlargement debate saying they did not want to have a second Greek-Turkish case in NATO. The official maintained that the experiences of the Greek-Turkish case still haunted officials. He asserted that the Alliance would be finished soon if it started to interfere in inter-member disputes more actively and cited the wisdom formulated by an author, whose name he could not remember, saying that a clever alliance knew when to retrench. Overall, there does not seem to be any exaggerated expectations and the officials appear to share the view that other factors have a greater impact on relations among aspirant countries. In this context, the NATO official mentioned just above pointed to the cases of Poland and Russia. He maintained that since both states wanted to improve ties, the fact that Poland joined NATO in 1999 did not have any repercussions on their relations (Interviews in Brussels, July 2001). The point here is that NATO officials are well aware of the fact that the Alliance cannot resolve conflicts among members. The strategy employed during the enlargement process has thus been to move aspirant countries to resolve any outstanding issues before they join the organization. As is the case with democracy, NATO seems to be able to make its greatest contribution while these countries are still outside knocking on its door.[119] Yet, the *conditionality of membership* is no guarantee that bilateral

---

[119]Here the question arises whether the conditions in Eastern Europe had really been as bad as assumed in the initial post-Cold War era. Reiter points out that despite continuous concerns that ethnic or territorial disputes might drive the region into chaos, the only clashes have occurred in the former Republic of Yugoslavia (and in the faraway Caucasus). Thus he argues that the prospect of NATO membership as a means for facilitating the peaceful resolution of conflict or pacifying interstate relations has not really been required (cf. 2001, 49-50). Furthermore, in his view, the often-cited example of reconciliation between Hungary and Romania cannot be solely attributed to the prospect of NATO membership. Domestic changes and the desire to join the EU had at least been equally important. Apart from that, the question to be asked is why the same mechanism did not bring about a reconciliation between Romania and countries like Moldova, Ukraine, or Russia. Reiter adds that above all the experience of Greek-Turkish animosity jeopardizes the assumption made about the pacifying effects of NATO membership. He holds that

conflicts will not be revived in the future. As a NATO official cautioned, aspirant countries might freeze their conflicts in order to be able to join the Alliance and defreeze them after having been admitted.

NATO membership has obviously fallen short of bringing about a Greek-Turkish reconciliation. However, this is not to say that the Greek-Turkish case refutes the assumptions formulated by institutionalists. It has usually been the clear issuance of the threat of force rather than the misperception of such a threat that has caused tensions in their relations. A lack of mutual trust and the *burden of history* have, of course, exacerbated their threat perceptions and complicated the situation. However, it has not been *imperfect conditions* - that is to say, insufficient information, misperceptions, or transaction costs (etc.) - constituting the problem in the first place, but much more conflicting interests and objectives. Moreover, it is hard to conceive how an institution can contribute to the resolution of a conflict where one of the parties has, at times, even declined to enter into dialogue with the opponent, viewing such a move as a concession. In sum, their differences have obviously outweighed the incentives for a settlement, and NATO seems to have been unable to provide the *carrots* that would have made reconciliation appear more profitable.

In comparison to NATO, though, the EU seems to be able to provide such *carrots* and contribute to an improvement in Greek-Turkish relations. Even though the reasons for the rapprochement process between the two countries seems to have been initiated by a bundle of factors as outlined above, it would not be wrong to say that the prospect of membership in the EU and the expectation that eventual membership will entail better economic, political, and social conditions have moved Turkey to upgrade its efforts towards the solution of the Cyprus issue and the outstanding matters in the Aegean. Meanwhile, Greek strategy seems to have changed as well. Before the Helsinki summit of 1999, Greeks preferred to bloc Turkey's way into the EU and keep the neighbor at distance with the Union. However, Greece realized that this strategy was contra-productive and simply exacerbated the situation. This obviously led the Greek decision-makers to undertake efforts to bring Turkey closer to Europe and support its EU bid. The Greeks must have understood that only the real perspective of membership could make Ankara accept certain concessions, simultaneously enabling itself to refashion Greek-Turkish disputes as EU-Turkish ones. Greek Foreign Minister Molyviatis' suggestion that Greek support for Turkish membership was not rooted in emotions but constituted the result of deep-rooted strategic considerations

---

"NATO was famously unsuccessful in preventing conflict between Greece and Turkey during the Cold War" (2001, 50).

(Hürriyet, September 22nd, 2004) is exemplary for the Greek approach of recent years. It is also worth recalling that Greece faced a necessity to cut military spending in order to comply with the EMU criteria. Overall, as long as both parties are willing to put an end to their centuries old conflict, be it in an attempt to cut defense expenditures or for any other reason, and show goodwill, the *EU factor* is likely to continue to act as a catalyst.

This is of course not to say that the problems will be solved automatically in the coming years if Turkey's membership prospect is kept upright. According to newspaper reports, as of late 2004, only modest progress has been made within the framework of the exploratory talks that have been going on for more than two years now (TDN, October 21st, 2004). Even though when granting the candidate-status to Turkey in 1999 the Union called on both parties to refer their disputes to the International Court of Justice in The Hague should they not be able to find a solution until the end of 2004, it is not clear yet whether this will really be the case. Former Turkish Ambassador to Greece Alpoğan stated in September that the disputes between the two countries had a technical content which required these negotiations to be conducted without any pressure of time (TDN, September 7th, 2004). The *Turkish Daily News* also reported Premier Erdoğan as saying that he was committed to resolving the disputes with Greece, yet these efforts should not be bound by any timeframe since the issues dealt with were complex and interrelated (October 25th, 2004). It is also worth remembering that Turkey has declared it would not recognize Cyprus prior to the commencement of negotiation talks. Thus the prospect of EU membership should not be expected to wipe away all remaining differences between the two countries. Moreover, the EU strategy seems to have one major "drawback." The Union allowed the Greeks and later Greek Cypriots - mainly under pressure from the Greeks - to enter the EU despite the problems with Turkey. At the same time, on December 17th, Turkey is likely to be asked to recognize Cyprus and commit itself to the resolution of border disputes, as if these disputes were only one-sided. Turkish Premier Erdoğan pointed to this ambiguity when he noted, "I am asking now: There is a condition among the EU membership criteria saying there should not be any border problems. How could Cyprus be granted membership even though it had border problems and had not solved these problems?" (Hürriyet, December 1st, 2004).[120] In comparison to Turkey, Greek Cypriots were allowed to join the Union on May 1st although they had voted against the UN plan that would have ended the division of the island. The Cypriot President had even called on the population to vote against the plan, reminding that they would join the EU anyway. Greeks and Greek Cypriots are likely to exert pressure on Turkey on December 17th or in the following period to make painful concessions. However, given the fact that

---

[120] Translated by the author.

according to the draft document leaked to the press Turkey is going to be confronted with harsh conditions, - for example, that negotiations can be suspended, permanent limitations can be put on the freedom of movement, or that accession talks need not necessarily lead to membership, an EU-Greek-Cypriot strategy to play for maximum gains could have adverse effects, stripping the EU of its role as a catalyst for solution. One wonders also whether the *Cypriot obstacle* could serve as a welcome pretext for those forces objecting Turkish membership to say "no" to the country in the end. For the time being, it is hard to imagine how the EU will handle this issue and how the relations among the parties involved will evolve. What seems to be unquestionable is that the *EU factor*, which has been used in a constructive manner since the 1999 Helsinki summit and contributed to an improvement in relations, can also turn into another issue of discontent burdening the relations between Greece and Turkey, as had been the case in the period preceding Helsinki. What is more, even if the outcome of this process was to be Turkish membership and the settling of all disputes, EU membership should not be viewed as a guarantee against the re-emergence of the conflict in the future. Especially the conflict on Cyprus, which is an ethnic conflict as such, can surface again long after all parties have acceded to the EU. The examples of Northern Ireland and the Basques in Spain show that membership in the EU is no guarantee against ethnic conflict. This is not to say that Turkey should not be admitted but that a steadfast comprehensive settlement is required meeting the expectations and winning the acceptance of all parties.

In contrast to the experience made in NATO, the Turkey-EU relationship also shows that international institutions might very well play an important role in the democratization process of a country. As shown before, Turkey has undertaken real efforts in recent years to bring its legal system in line with European standards, mainly in an attempt to join the Union. The European Union, on the other hand, has been monitoring the situation in Turkey very closely. In the end, the report presented by the Commission concluded that Turkey had met the political criteria and recommended the opening of talks. However, Turkey's legal system and democratic record will continue to be scrutinized by the Union. The Commission suggested that the accession talks be suspended should Turkey leave the path of reform or should the Union identify serious breaches of democratic principles or human rights. Thus Turkish authorities have no option but to implement and adhere to the reforms of recent years if they want to join the Union at any time in the future - or even if it turns out in 10 years time that Turkey will not be accepted as a member. On the other hand, leaving Turkey out in the cold is likely to have its repercussions on the democratization process there. In this sense, EU's former Commissioner for enlargement and the mastermind of the October 6th report, Günther Verheugen, even claimed that a risk involved in saying no to

Turkey might be that political reform comes to an end or even fails (IHT, October 7th, 2004). Even though the opinion offered by Verheugen might be exaggerated, as the reform process in Turkey should not be solely attributed to the prospect of EU membership, it is unquestionable that the EU has acted as a catalyst and made a difference with regard to the extent and pace of reforms.

Returning to the theoretical dimension of the debate, the fact that NATO membership has had ambivalent effects on the course of relations between Greece and Turkey and has fallen short of promoting genuine democratic reform, while the EU, simultaneously, has recently had more success in both fields attests to the suggestion made by institutionalists that institutional effects vary in different settings. NATO has usually found itself in a position where two states have been in conflict and had no incentives to solve it. Under such conditions, institutionalists would not expect institutions to matter much. The EU, however, has been able to offer certain *carrots* making a resolution appear more attractive to two countries which seem to be willing to put aside their differences and contribute to the stability in their surroundings rather than acting as "Europe's spoilt kids." Thus, the Greek-Turkish case shows that international institutions do matter, even though different institutions might unfold different effects and these effects need not always be positive. In the end, one should not forget that institutions constitute only one among various factors influencing the behavior of states, and that the extent and efficacy of the role played by this factor varies in different settings. While this role might sometimes be more prominent than in other cases, ultimately, as Keohane puts it "in modern international relations, the pressures from domestic interests, and those generated by the competitiveness of the state system, exert much stronger effects on state policy than do international institutions, even broadly defined" (Keohane 1989, 6).

145

BIBLIOGRAPHY

Abbott W. Kenneth & Duncan Snidal: *Why States Act through Formal International Organizations*, in: Journal of Conflict Research, Vol. 42, No. 1 (February 1998), pp. 3-32.

Alford Jonathan: *Greece and Turkey: Adversity in Alliance*, Aldershot 1984.

Altunışık Meliha: *The Political Economy of Caspian Oil*, in: Erich Reiter (ed.), Jahrbuch für internationale Sicherheitspolitik 1999, Hamburg 1999, pp. 674-684.

Art J. Robert: *Creating a Disaster: NATO`s Open Door Policy*, in: Political Science Quarterly, Vol. 113, No. 3, 1998, pp. 383- 403.

Asmus D. Ronald, Richard L. Kugler & Stephen Larrabee: *Building a New NATO*, in: Foreign Affairs, Vol. 72, No. 4, September/October 1993, pp. 28-40.

Auernheimer Gustav: *Zum Bild der Türkei in Griechenland und seinen historischen Voraussetzungen*, in: Südosteuropa, Vol. 48, No. 5-6, 1999, pp. 336-358.

Auernheimer Gustav: *Der Kosovokonflikt und die griechische Öffentlichkeit*, in:Südosteuropa, Vol. 48, No. 7-8, 1999, pp. 389-400.

Auernheimer Gustav: *Griechenland vor dem Beitritt zur Europäischen Währungsunion*, in: Südosteuropa, Vol. 49, No. 3-4, 2000, pp. 144-156.

Axelrod Robert & Robert O. Keohane: *Achieving Cooperation under Anarchy: Strategies and Institutions*, in: Baldwin A. David (ed.), Neorealism and Neoliberalism: The Contemporary Debate, New York 1993, pp. 85-113.

Axt Heinz-Jürgen: *Auf dem Balkan isoliert. Griechenlands außen- und sicherheitspolitische Situation nach dem Ende des Ost- West- Konfliktes*, in: Südosteuropa Mitteilungen, Vol. 33, No. 3, 1993, pp. 242-251.

Axt Heinz-Jürgen: *Zypern – ein Beitrittskandidat der Europäischen Union. Implikationen für die Insel, die Region und die Union*, in: Südosteuropa, Vol. 44, No. 5, 1995, pp. 259-279.

Axt Heinz-Jürgen: *Der Ägäis-Streit – ein unlösbarer griechisch-türkischer Konflikt?*, in: Südosteuropa Mitteilungen, Vol. 39, No. 2, 1999a, pp. 137-151.

Axt Heinz-Jürgen: *Zypern und der „Acquis politique": Außen- und Sicherheitspolitik in der Perspektive des EU-Beitritts einer geteilten Insel*, in: Südosteuropa Mitteilungen, Vol. 39, No. 4, 1999b, 319-333.

Axt Heinz-Jürgen: *Selbstbewusstere Türkei. Worauf sich die EU einstellen muss*, in: Internationale Politik, 1/2001, pp. 45-50.

Axt Heinz-Jürgen: *Gordischer Knoten in Kopenhagen nicht durchschlagen: Zypern, die Türkei und die EU*, in: Integration, Vol. 26, No. 1, 2003, pp. 66-77.

Axt Heinz-Jürgen & Heinz Kramer: *Entspannungspolitik im Ägäiskonflikt? – Griechisch-türkische Beziehungen nach Davos*, Baden-Baden 1990.

146

Ayman S. Gülden: *Türk – Yunan İlişkilerinde Güc ve Tehdit*, in: Faruk Sönmezoğlu (ed.), Türk Dış Politikasının Analizi, İstanbul 2001, pp. 543-555.

Bağcı Hüseyin: *Die Zypernpolitik der Menderes-Regierung von 1950-1960. Ein Wendepunkt in der türkischen Außenpolitik*, in: Orient, Vol. 33, No. 1, 1992, pp. 119-138.

Bağcı Hüseyin: *Türkische Sicherheitspolitik*, in: Internationale Politik, 1/1998, pp. 29-34.

Bağcı Hüseyin: *Türkische Außenpolitik nach dem Luxemburger EU-Gipfel vom Dezember 1997: Europäisch ohne Europa?*, in: Erich Reiter (ed.), Jahrbuch für internationale Sicherheitspolitik 1999, Hamburg 1999, pp. 579- 602.

Bahçeli Tozun: *The Potential for Conflicts in Greek-Turkish Relations*, in: Erich Reiter (ed.), Jahrbuch für internationale Sicherheitspolitik 2000, Hamburg 2000, pp.457-471.

Bahçeli Tozun: *Turkey`s Cyprus Challenge. Preserving the Gains of 1974*, in: Dimitris Keridis, Dimitirios Triantaphyllou (eds.), Greek – Turkish Relations in the Era of Globalization, Everett 2001, pp. 208-222.

Baldwin A. David: *Neoliberalism, Neorealism, and World Politics*, in: Baldwin A. David (ed.), Neorealism and Neoliberalism: The Contemporary Debate, New York 1993, pp. 4-25.

Ball George: *The Past has another Pattern*, New York 1982.

Balmaceda M. Margarita: *Institutions, Alliances and Stability: Thinking Theoretically about International Relations in Central–East Europe*, in: European Security, Vol. 6, No. 3, Autumn 1997, pp. 85-109.

Barnett N. Michael & Jack S. Levy: *Domestic Sources of Alliances: the Case of Egypt, 1962-73*, in: International Organization, Vol. 45, No. 3, Summer 1991, pp. 369-395.

Baytok Taner: *Bir Asker Bir Diplomat. Güven Erkaya – Taner Baytok. Söyleşi*, İstanbul 2001.

Birand A. Mehmet: *30 Sıcak Gün*, İstanbul 1976.

Birand A. Mehmet: *Turkey and the `Davos Process`: Experiences and Prospects*, in: Dimitri Constas (ed.), The Greek – Turkish Conflict in the 1990s. Domestic and External Influences, London 1991, pp. 27-39.

Boll M. Michael: *Turkey`s New National Security Concept: What it means for NATO*, in: ORBIS, Fall 1979, pp. 609-622.

Bölükbaşı Suha: *The Turco-Greek Dispute: Issues, Policies and Prospects*, in: Clement H. Dodd (ed.), Turkish Foreign Policy: New Prospects, Occasional Papers, No.2, Huntingdon 1992, pp. 27-51.

147

Bozo Frederic: *Continuity or Change? The View from Europe*, in: Victor S. Papacosma, Sean Kay, Mark R. Rubin (eds.), NATO after Fifty Years, Wilmington 2001, pp. 53-72.

Brosio Manlio: *Consultation and the Atlantic Alliance*, in: The Atlantic Community Quarterly, Vol. 12, No. 3, Fall 1974, pp. 308-318.

Bulaç Ali: *Avrupa Birliği ve Türkiye*, İstanbul 2001.

Çağrı Erhan: *ABD ve NATO'yla İlişkiler* [1945-1960], in: Baskın Oran (ed.), Türk Dış Politikası. Kurtuluş Savaşından Bugüne Olgular, Belgeler, Yorumlar, Volume I, İstanbul 2001a, pp. 522-575.

Çağrı Erhan: *ABD ve NATO'yla İlişkiler* [1960-1980], in: Baskın Oran (ed.), Türk Dış Politikası. Kurtuluş Savaşından Bugüne Olgular, Belgeler, Yorumlar, Volume I, İstanbul 2001b, pp.681-715.

Callaghan James: *Time and Change*, London 1987.

Caporaso A. James: *International Relations Theory and Multilateralism: The Search for Foundations*, in: International Organization, Vol. 46, No. 3, Summer 1992, pp. 599-632.

Çarkoğlu Ali & Mine Eder: *Water Conflict: The Euphrates – Tigris Basin*, in: Barry Rubin, Kemal Kirişci (eds.), Turkey in World Politics. An Emerging Multiregional Power, Boulder 2001, pp. 241-243.

Carpenter G. Ted: *NATO's Search for Relevance*, in: Victor S. Papacosma, Sean Kay, Mark R. Rubin (eds.), NATO after Fifty Years, Wilmington 2001, pp. 25 – 41.

Carr Fergus & Kostas Ifantis: *NATO in the New European Order*, New York 1996.

Carrington A. R. Peter: *Reflect on Things Past. The Memoirs of Lord Carrington*, London 1988.

Cartier Claude: *NATO and the European Union*, in: Victor S. Papacosma, Mary Ann Heiss (eds.), NATO in the Post-Cold War Era: Does it have a Future?, Houndsmill 1993, pp. 136 – 150.

Celeste A. Wallander/ Robert O. Keohane: *Risk, Threat, and Security Institutions*, in: Helga Haftendorn, Robert O. Keohane, Celeste A. Wallander (eds.), Imperfect Unions – Security Institutions over Time and Space, Oxford 1999, pp. 21-47.

Chase S. Robert, Emily B. Hill & Paul Kennedy: *Pivotal States in US Strategy*, in: Foreign Affairs, Vol. 75, No. 1, January/February 1996, pp.33-51.

Chipman John (ed.): *NATO's Southern Allies: Internal and External Challenges*, London 1988a.

Chipman John: *NATO and the Security Problems of the Southern Region: From the Azores to Ardahan*, in: John Chipman (ed.), NATO's Southern Allies: Internal and External Challenges, London 1988b, pp. 8-52.

Chipman John: *Allies in the Mediterranean: Legacy of Fragmentation*, in: John Chipman (ed.), NATO`s Southern Allies: Internal and External Challenges, London 1988c, pp. 53-85.

Chipman John: *NATO`s Southern Region: National versus Alliance Priorities*, in: John Chipman (ed.), NATO`s Southern Allies: Internal and External Challenges, London 1988d, pp. 354-381.

Clogg Richard: *Griechenland und die Zypernkrise*, in: Europa-Archiv, 18/1974, pp. 633-642.

Clogg Richard: *Greek – Turkish Relations in the post – 1974 Period*, in: Dimitri Constas (ed.), The Greek – Turkish Conflict in the 1990s. Domestic and External Influences, London 1991, pp. 12-23.

Clogg Richard: *A Concise History of Greece*, Cambridge 20002.

Commission of the European Communities: *2004 Regular Report on Turkey's Progress towards Accession*, October 2004a, http://www.eu.int/ comm/enlargement/report_2004/pdf/rr_tr_2004_en.pdf.

Commission of the European Communities: *Issues Arising from Turkey's Membership Perspective*, October 2004b, http://www.europa.eu.int/comm/ enlargement/ report_2004/pdf/issues_paper_en.pdf.

Commission of the European Communities: *Recommendation of the European Commission on Turkey's Progress towards Accession*, October 2004c, http://www.eu.int/comm/enlargement/report_2004/pdf/tr_recommandation _en.pdf.

Constas Dimitri (ed.): *The Greek-Turkish Conflict in the 1990s. Domestic and External Influences*, London 1991.

Constas Dimitri: *Systemic Influence on a Weak, Aligned State in the post – 1974 Era*, in: Dimitri Constas (ed.), The Greek – Turkish Conflict in the 1990s. Domestic and External Influences, London 1991, pp. 129-139.

Coufoudakis Van: *Greek-Turkish Relations, 1973-1983*, in: International Security, Vol. 9, No. 4, Spring 1985, pp. 185-217.

Coufoudakis Van: *Greek Foreign Policy since 1974: Quest for Independence*, Journal of Modern Greek Studies, 6, 1988, 55-79.

Coufoudakis Van: *Greek Political Party Attitudes towards Turkey: 1974-89*, in: Dimitri Constas (ed.), The Greek – Turkish Conflict in the 1990s. Domestic and External Influences, London 1991.

Coufoudakis Van: *PASOK on Greco-Turkish Relations and Cyprus, 1981-1989: Ideology, Pragmatism, Deadlock*, in: Theodore C. Kariots (ed.), The Greek Socialist Experiment-Papandreou`s Greece 1981-1989, New York 1992.

Cutler Robert: *Soviet Relations with Greece and Turkey: A Systems Perspective*, in: Dimitri Constas (ed.), The Greek – Turkish Conflict in the 1990s. Domestic and External Influences, London 1991, pp. 183-206.

149

Dedeoğlu Beril: *Değişen Uluslararası Sistemde Türkiye – ABD İlişkilerinin Türkiye – Avrupa Birliği İlişkilerine Etkileri*, in: Faruk Sönmezoğlu (ed.), Türk Dış Politikasinin Analizi, İstanbul 2001, pp.227-251.

Demirbaş-Coşkun Birgül: *Ankara–Atina Hattında Son Gelişmeler: "Türk–Yunan Baharı"nın Sonu mu?*, in: Stratejik Analiz, Vol. 1, No. 8, December 2000, pp.26-30.

Deringil Selim: *Turkish Foreign Policy since Atatürk*, in: Clement H. Dodd (ed.), Turkish Foreign Policy: New Prospects, Occasional Papers, No. 2, Huntingdon 1992, pp. 1-6.

Douglas T. Stuart (ed.): *Politics in the Southern Region of the Atlantic Alliance*, Houndsmill 1988.

Duffield S. John: *International Regimes and Alliance Behavior: Explaining NATO Conventional Force Levels*, in: International Organization, Vol. 46, No. 4, Autumn 1992, pp. 819-855.

Duffield S. John: *NATO`s Functions after the Cold War*, in: Political Science Quarterly, Vol. 109, No. 5, 1994-95, pp. 763- 787.

Duke W. Simon: *Small States and European Security*, in: Erich Reiter, Heinz Gärtner (eds.), Small States and Alliances, Heidelberg 2001, pp. 39-50.

Eder Mine: *The Challenge of Globalization and Turkey`s Changing Political Economy*, in: Barry Rubin, Kemal Kirişci (eds.), Turkey in World Politics. An Emerging Multiregional Power, Boulder 2001, pp. 189-194.

Eren Nuri: *Turkey, NATO and Europe: a Deteriorating Relationship?*, Atlantic Papers, No. 34, Paris 1977.

Ergüvenç Sadi: *Turkey`s Security Perceptions*, in: Perceptions – Journal of International Affairs, Vol. 3, No. 2, June-August 1998, pp. 32-42.

Ergüvenç Sadi: *Ege Denizinde Karasuları Anlaşmazlığı – "Gereksiz Bir Krizin Ardından"*, in: Strateji, 95/1, pp. 5-20.

Evangelista Matthew: *Greece, Turkey and the Improvement of US- Soviet Relations*, in: Dimitri Constas (ed.), The Greek – Turkish Conflict in the 1990s. Domestic and External Influences, London 1991, pp. 140-154.

Fırat Melek: *Yunanistan'la İlişkiler* [1923-1939], in: Baskın Oran (ed.), Türk Dış Politikası. Kurtuluş Savaşından Bugüne Olgular, Belgeler, Yorumlar, Volume I, İstanbul 2001a, pp. 325-356.

Fırat Melek: *Yunanistan'la İlişkiler* [1945-1960], in: Baskın Oran (ed.), Türk Dış Politikası. Kurtuluş Savaşından Bugüne Olgular, Belgeler, Yorumlar, Volume I, İstanbul 2001b, pp.576-614.

Fırat Melek: *Yunanistan'la İlişkiler* [1960-1980], in: Baskın Oran (ed.), Türk Dış Politikası. Kurtuluş Savaşından Bugüne Olgular, Belgeler, Yorumlar, Volume I, İstanbul 2001c, pp. 716-767.

Fırat Melek: *Yunanistan'la İlişkiler* [1980-1990], in: Baskın Oran (ed.), Türk Dış Politikası. Kurtuluş Savaşından Bugüne Olgular, Belgeler, Yorumlar, Volume II, İstanbul 2001d, pp. 102-123.

Fırat Melek: *Yunanistan'la İlişkiler* [1990-2001], in: Baskın Oran (ed.), Türk Dış Politikası. Kurtuluş Savaşından Bugüne Olgular, Belgeler, Yorumlar, Volume II, İstanbul 2001e, pp. 440-480.

Forndran Erhard: *Die Reichweite der Modelle der Friedensforschung in der Politischen Praxis: Das Beispiel des griechisch-türkischen Konfliktes*, in: Orient, Vol. 27, No. 3, September 1986, pp. 409-422.

Gärtner Heinz: *Modelle Europäischer Sicherheit. Wie entscheidet Österreich?*, Wien 1997.

Gärtner Heinz & Allen G. Sens: *The Adaptability of Institutions and Small States*, in: Ingo Peters (ed.), New Security Challenges. The Adaptation of International Institutions; Reforming the UN, NATO, EU and CSCE since 1989, Münster 1996, pp. 179-201.

Gelpi Christopher: *Alliances as Instruments of Intra-Allied Control*, in: Helga Haftendorn, Robert O. Keohane, Celeste A. Wallander (eds.), Imperfect Unions – Security Institutions over Time and Space, Oxford 1999, pp. 107-138.

Giakoumis Pantelis: *Die Macht als Faktor in der Außenpolitik Griechenlands und der Türkei*, in: Südosteuropa, Vol. 40, No. 9, 1991, pp. 441-472.

Gilpin Robert: *The Theory of Hegemonic War*, in: Robert I. Rotberg, Theodore K. Rabb, The Origins and Prevention of Major Wars, Cambridge 1988, pp. 15 – 34.

Glaser L. Charles: *Why NATO is Still Best. Future Security Arrangements for Europe*, in: International Security, Vol. 18, No. 1, Summer 1993, pp. 5-50.

Glaser L. Charles: *Realists as Optimists. Cooperation as Self-Help*, in: International Security, Vol. 19, No. 3, 1994795, pp. 50-90.

Glaser L. Charles: *The Security Dilema Revisited*, in: World Politics, Vol. 50, No. 1, October 1997, pp. 171-201.

Gowa Joanne: *Rational Hegemons, Excludable Goods, and Small Groups: An Epitaph for Hegemonic Stability Theory?*, in: World Politics, Vol. 41, No. 3, 1989, pp. 307-324.

Grieceo M. Joseph: *Anarchy and the Limits of Cooperation: A Realist Critique of the Newest Liberal Institutionalism*, in: Baldwin A. David (ed.), Neorealism and Neoliberalism: The Contemporary Debate, New York 1993a, pp. 116-131.

Grieco M. Joseph: *Understanding the Problem of International Cooperation. The Limits of Neoliberal Institutionalism and the Future of Realist Theory*, in: Baldwin A. David (ed.), Neorealism and Neoliberalism: The Contemporary Debate, New York 1993b, pp. 301-332.

Güldemir Ufuk: *Kanat Operasyonu*, İstanbul 1986.

Gumpel Werner: *Determinanten der türkischen Außenpolitik in der Schwarzmeerregion und in Mittelasien*, in: Südosteuropa Mitteilungen, Vol. 38, No. 1, 1998, pp. 23-32.

Gumpel Werner: *Ordnungsfaktor in unsicherem Umfeld*, in: Internationale Politik, 11/2000, pp. 21-27.

Gündüz Aslan: *Greek – Turkish Disputes. How to Resolve Them*, in: Dimitris Keridis, Dimitirios Triantaphyllou (eds.), Greek – Turkish Relations in the Era of Globalization, Everett 2001, pp. 81-101.

Gürbey Gülüstan: *Die türkisch-syrische Krise: Nur eine Kriegsdrohung?*, in: Südosteuropa Mitteilungen, Vol. 38, No. 4, 1998, pp. 349-359.

Gürbey Gülüstan: *Der Fall Öcalan und die türkisch-griechsiche Krise: Alte Drohungen oder neue Eskalation*, in: Südosteuropa Mitteilungen, Vol. 39, No. 2, 1999, pp.123-136.

Gürkan İhsan: *1974 Kıbrıs Barış Harekatı`nda Siyasal İradenin Oluşumu ve Askeri Uygulama*, in: Faruk Sönmezoğlu (ed.), Türk Dış Politikasının Analizi, İstanbul 2001, pp. 271-279.

Gürün Kamuran: *Türk Yunan İlişkileri – İtimat Unsuru*, in: Dış Politika, Vol. 10, No. 1, 1977, pp. 30-33.

Haass N. Richard: *Managing NATO`s Weakest Flank: The United States, Greece, and Turkey*, in: ORBIS, Fall 1986, pp. 457-473.

Haass Richard: *Alliance Problems in the Eastern Mediterranean – Greece, Turkey and Cyprus: Part I*, in: IISS, Prospects for Security in the Mediterranean, Adelphi Papers, No. 229, London 1988, pp. 61-71.

Haftendorn Helga: *Sicherheitsinstitutionen in den internationalen Beziehungen. Eine Einführung*, in: Helga Haftendorn, Otto Keck (eds.), Kooperation jenseits von Hegemonie und Bedrohung. Sicherheitsinstitutionen in den internationalen Beziehungen, Baden-Baden 1997, pp.11-33.

Haftendorn Helga: *The 'Quad': Dynamics of Institutional Change*, in: Helga Haftendorn, Robert O. Keohane, Celeste A. Wallander (eds.), Imperfect Unions – Security Institutions over Time and Space, Oxford 1999, pp. 162-194.

Haftendorn Helga: *Das Ende der alten NATO*, in: Internationale Politik, 4/2002, pp. 49-54.

Haftendorn Helga/ Robert O. Keohane/ Celeste A. Wallander (eds.): *Imperfect Unions – Security Institutions over Time and Space*, Oxford 1999.

Hale William & Gamze Avci: *Turkey and the European Union: The Long Road to Membership*, in: Barry Rubin, Kemal Kirişci (eds.), Turkey in World Politics. An Emerging Multiregional Power, Boulder 2001, pp. 31-44.

152

Harris S. George: *Troubled Alliance: Turkish – American Problems in Historical Perspective*, 1945-1971, Washington, D.C. 1972.

Hart T. Parker: *Two NATO Allies at the Threshold of War: Cyprus, a first Account of Crisis Management 1965-1968*, Durham 1990.

Hasenclever Andreas: *The Democratic Peace meets International Institutions*, in: Zeitschrift für Internationale Beziehungen, Vo. 9, No. 1, 2002, pp. 75-112.

Heiss A. Mary: *NATO and the Middle East: The Primacy of National Interest*, in: S. Victor Papacosma & Mary Ann Heiss, NATO in the post-Cold War Era: Does it have a Future?, Houndsmill 1995, pp. 279-297.

Hellmann Gunther & Reinhard Wolf: *Neorealism, Neoliberal Institutionalism, and the Future of NATO*, in: Security Studies, Vol. 3, No. 1, Autumn1993, pp. 3-43.

Hickok R. Michael: *The Imia/Kardak Affair, 1995-96: A Case of Inadvertent Conflict*, in: European Security, Vol. 7, No. 4, Winter 1998, pp. 118-136.

Hixson L. Walter: *NATO and the Soviet Bloc: The Limits of Victory*, in: S. Victor Papacosma & Mary Ann Heiss, NATO in the post-Cold War Era: Does it have a Future?, Houndsmill 1995, pp. 23-38.

Hohenecker Henrike: *Die Gründungsphase der NATO 1945-1955. Wie wurde die NATO ein Militärbündnis?*, Masters Thesis, University of Vienna, 2000.

Hunter Shireen: *Bridge or Frontier? Turkey's Post-Cold War Geopolitical Posture*, in: The International Spectator, Vol. 34, No. 1, January – March 1999, pp. 63-78.

Iatrides O. John: *Papandreou`s Foreign Policy*, in: Theodore C. Kariots (ed.), The Greek Socialist Experiment-Papandreou`s Greece 1981-1989, New York 1992.

Iatrides O. John: *Greece in the Cold War, and Beyond*, in: Journal of Hellenic Diaspora, 19, 1993, pp.11-30.

Ierodiakonou Leontios: *The Cyprus Question*, Stockholm 1971.

Inbar Efraim: *The Strategic Glue in the Israeli – Turkish Alignment*, in: Barry Rubin, Kemal Kirişci (eds.), Turkey in World Politics. An Emerging Multiregional Power, Boulder 2001, pp. 115-126.

Independent Commission on Turkey: *Turkey in Europe: More than a Promise?*, September 2004, http://www.soros.org/resources/articles_publications/ publications/turkey_2004901.

International Institute for Strategic Studies: *Strategic Survey 1997/98*, London 1998.

Ireland P. Timothy: *Creating the Entangling Alliance – The Origins of the North Atlantic Treaty Organisation*, Westport 1981.

Jenkins Gareth: *Turkey's Changing Domestic Politics*, in: Dimitris Keridis & Dimitirios Triantaphyllou (eds.), Greek – Turkish Relations in the Era of Globalization, Everett 2001, pp. 19-40.

Jervis Robert: *Cooperation under the Security Dilemma*, in: World Politics, Vol. 30, No. 2, 1978, pp. 167-214.

Jervis Robert: *Security Regimes*, in: International Organization, Vol. 36, No. 2, Spring 1982, pp. 357-378.

Jervis Robert: *Realism, Neoliberalism, and Cooperation. Understanding the Debate*, in: International Security, Vol. 24, No. 1, Summer 1999, pp. 42-63.

Jordan S. Robert: *Political Leadership in NATO: A Study in Multinational Diplomacy*, Boulder 1979.

Jordan S. Robert: *NATO's Structural Changes for the 1990s*, in: Victor S. Papacosma, Mary Ann Heiss (eds.), NATO in the post-Cold War Era: Does it have a Future?, Houndsmill 1993, pp. 41 –54.

Jordan S. Robert: *NATO as a Political Organization. The Supreme Allied Commander, Europe (SACEUR) and the Politics of Command*, in: Victor S. Papacosma, Sean Kay, Mark R. Rubin (eds.), NATO after Fifty Years, Wilmington 2001, pp. 87-119.

Joseph S. Joseph: *Ethnic Loyalties vs. Allied Commitments: Greek-Turkish Conflict over Cyprus as a Source of Strain for NATO*, in: Thetis, 2/1995, pp. 235-243.

Kale Başak: *Türkiye'nin AB Serüveninde Yeni Bir Sayfa: Katılım Ortaklık Belgesi*, in: Stratejik Analiz, Vol. 1, No. 8, December 2000, pp. 5-23.

Kale Başak: *Ulusal Program ile Türkiye'nin AB için Değişen Statüsü: "Ebedi Adaylık"*, in: Stratejik Analiz, Vol. 2, No. 13, May 2001, pp. 20-24.

Kaminaris Ch. Spiros: *Greece and the Middle East*, in: MERIA, Vol. 3, No. 2, June 1999 (web edition: www.biu.ac.il).

Kaplan S. Lawrence: *NATO after Forty-Five Years: A Counterfactual History*, in: Victor S. Papacosma, Mary Ann Heiss (eds.), NATO in the Post-Cold War Era: Does it have a Future?, Houndsmill 1993, pp. 3 – 21.

Kaplan S. Lawrence: *The United States and NATO. The Relevance of History*, in: Victor S. Papacosma, Sean Kay, Mark R. Rubin (eds.), NATO after Fifty Years, Wilmington 2001, pp. 243-261.

Karaosmanoğlu Ali: *Turkey's Security Policy: Continuity and Change*, in: Douglas T. Stuart (ed.), Politics and Security in the Southern Region of the Atlantic Alliance, Houndsmill 1988a, pp. 157-180.

Karaosmanoğlu Ali: *Turkey and the Southern Flank: Domestic and External Contexts*, in: John Chipman (ed.), NATO's Southern Allies: Internal and External Challenges, London 1988b, pp. 287-353.

154

Karaosmanoğlu L. Ali: *The International Context of Democratic Transition in Turkey*, in: Geoffrey Pridham (ed.), Encouraging Democracy – The International Context of Regime Transition in Southern Europe, Leicester 1991, pp. 159-174.

Karaosmanoğlu L. Ali: *NATO Enlargement and the South. A Turkish Perspective*, in: Security Dialogue, Vol. 30, Nr. 2, 1999, pp. 213-224.

Karaosmanoğlu L. Ali: *The Evolution of the National Security Culture and the Military in Turkey*, in: Journal of International Affairs, Vol. 54, No. 1, Fall 2000, pp. 199-216.

Karpat H. Kemal: *Military Interventions: Army-Civilian Relations in Turkey before and after 1980*, in: Metin Heper, Ahmet Evin (eds.), State, Democracy and the Military. Turkey in the 1980s, Berlin 1988, pp. 137-154.

Kay Sean: *NATO and the Future of European Security*, Lanham 1998.

Keck Otto: *Der Beitrag rationaler Theorieansätze zur Analyse von Sicherheitsinstitutionen*, in: Helga Haftendorn, Otto Keck (eds.), Kooperation jenseits von Hegemonie und Bedrohung. Sicherheits-institutionen in den internationalen Beziehungen, Baden-Baden 1997a, pp. 35-56.

Keck Otto: *Sicherheitsinstitutionen im Wandel des internationalen Systems*, in: Helga Haftendorn, Otto Keck (eds.), Kooperation jenseits von Hegemonie und Bedrohung. Sicherheitsinstitutionen in den internationalen Beziehungen, Baden-Baden 1997b, pp.253-270.

Keohane O. Robert: *The Demand for International Regimes*, in: International Organization, Vol. 36, No. 2, Spring 1982, pp.325-355.

Keohane O. Robert: *After Hegemony. Cooperation and Discord in the World Political Economy*, Princeton 1984.

Keohane O. Robert: *Alliances, Threats, and the Uses of Neorealism*, in: International Security, Vol. 13, No. 1, Summer 1988, pp. 169-176.

Keohane O. Robert: *International Institutions and State Power: Essays in Interanational Relations Theory*, Boulder 1989.

Keohane O. Robert: *Institutional Theory and the Realist Challenge after the Cold War*, in: Baldwin A. David (ed.), Neorealism and Neoliberalism: The Contemporary Debate, New York 1993, pp. 269-296.

Keohane O. Robert & Lisa L. Martin: *The Promise of Institutionalist Theory*, in: International Security, Vol. 20, No. 1, Summer 1995, pp. 39-51.

Keridis Dimitris: *Domestic Developments and Foreign Policy. Greek Policy toward Turkey*, in: Dimitris Keridis & Dimitirios Triantaphyllou (eds.), Greek – Turkish Relations in the Era of Globalization, Everett 2001, pp. 2-18.

Keridis Dimitris & Dimitirios Triantaphyllou: *Greek – Turkish Relations in the Era of European Integration and Globalization*, in: Dimitris Keridis & Dimitirios Triantaphyllou (eds.), Greek – Turkish Relations in the Era of Globalization, Everett 2001, pp. xvi-xxii.

Khrushcheva L. Nina: *Russia and NATO. Lessons Learned*, in: Victor S. Papacosma, Sean Kay, Mark R. Rubin (eds.), NATO after Fifty Years, Wilmington 2001, pp. 229-241.

Kirişci Kemal: *The End of the Cold War and Changes in Turkish Foreign Policy Behaviour*, in: Dış Politika – Foreign Policy, Vol. 17, No. 3-4, 1993, pp. 1-43.

Kirişci Kemal: *The Future of Turkish Policy toward the Middle East*, in: Barry Rubin, Kemal Kirişci (eds.), Turkey in World Politics. An Emerging Multiregional Power, Boulder 2001a, pp. 93-108.

Kirişci Kemal: *US – Turkish Relations: New Uncertainties in a Renewed Partnership*, in: Barry Rubin, Kemal Kirişci (eds.),Turkey in World Politics. An Emerging Multiregional Power, Boulder 2001b, pp. 129-145.

Kissinger Henry: *Years of Renewal*, New York 1999.

Klauer Vera: *Bedingungen institutioneller Leistungsfähigkeit am Beispiel des Konfliktes im ehemaligen Jugoslawien*, in: Helga Haftendorn, Otto Keck (eds.), Kooperation jenseits von Hegemonie und Bedrohung. Sicherheitsinstitutionen in den internationalen Beziehungen, Baden-Baden 1997, pp. 233-252.

Kourvetaris A. Yorgos: *The Southern Flank of NATO: Political Dimensions of the Greco-Turkish Conflict since 1974*, in: East European Quarterly, Vol. 21, No. 4, January 1988, pp. 431-446.

Kramer Heinz: *Turkey's Relations with Greece: Motives and Interests*, in: Dimitri Constas (ed.), The Greek – Turkish Conflict in the 1990s. Domestic and External Influences, London 1991, pp. 57-72.

Kramer Heinz: *A Changing Turkey: The Challenge to Europe and the United States*, Washington, D.C. 2000.

Krasner D. Stephen: *Structural Causes and Regime Consequences: Regimes as Intervening Variables*, in: International Organization, Vol. 36, No. 2, Spring 1982, pp. 185-205.

Krasner D. Stephen: *Global Communications and National Power: Life on the Pareto Frontier*, in: Baldwin A. David (ed.), Neorealism and Neoliberalism: The Contemporary Debate, New York 1993, pp. 234-247.

Krause Volker & David J. Singer: *Minor Powers, Alliances, and Armed Conflict: Some Preliminary Patterns*, in: Erich Reiter, Heinz Gärtner (eds.), Small States and Alliances, Heidelberg 2001, pp. 15-23.

Krebs R. Ronald: *Perverse Institutionalism: NATO and the Greco-Turkish Conflict*, in: International Organization, Vol. 53, No. 2, Spring 1999, pp. 343-377.

Kreft Michael: *Die Europäische Integration als Sicherheitsinstitution*, in: Helga Haftendorn, Otto Keck (eds.), Kooperation jenseits von Hegemonie und Bedrohung. Sicherheitsinstitutionen in den internationalen Beziehungen, Baden-Baden 1997, pp.165-190.

Kuniholm R. Bruce: *The Origins of the Cold War in the Near East – Great Power Conflict and Diplomacy in Iran, Turkey, and Greece*, Princeton 1980.

Kuniholm R. Bruce: *Turkey and NATO: Past, Present, and Future*, in: ORBIS, Vol. 27, No. 2, Summer 1983, pp. 421-445.

Kupchan A. Charles: *NATO and the Persian Gulf: Examining Intra-Alliance Behavior*, in: International Organization, Vol. 42, No. 2, Spring 1988, pp. 317-346.

Kupchan A. Charles: *Turning Adversity into Advantage: Russia in NATO*, in: Matthias Jopp & Hanna Ojanen (eds.), European Security Integration: Implications for Non-alignment and Alliances, Helsinki 1999.

Kürkcüoĝlu N. Elif: *Der türkisch-griechische Ägäiskonflikt in seiner historischen und völkerrechtlichen Dimension*, Masters Thesis, University of Vienna, 2000.

Kurop C. Marcia: *Greece and Turkey – Can They Mend Fences?*, in: Foreign Affairs, Vol. 77, No. 1, January/February 1998, pp. 7-12.

Kut Şule: *The Contours of Turkish Foreign Policy in the 1990s*, in: Barry Rubin, Kemal Kirişci (eds.), Turkey in World Politics. An Emerging Multiregional Power, Boulder 2001a, pp. 5-11.

Kut Şule: *Türk Dış Politikasında Ege Sorunu*, in: Faruk Sönmezoglu (ed.),Türk Dış Politikasının Analizi, İstanbul 2001b, pp. 253-270.

Laipson Helen: *US Policy towards Greece and Turkey since 1974*, in: Dimitri Constas (ed.), The Greek – Turkish Conflict in the 1990s. Domestic and External Influences, London 1991, pp. 164-182.

Larrabee F. Stephen: *Security in the Eastern Mediterranean. Transatlantic Challenges and Perspectives*, in: Dimitris Keridis, Dimitirios Triantaphyllou (eds.), Greek – Turkish Relations in the Era of Globalization, Everett 2001, pp. 224-238.

Laurent Pierre-Henri: *NATO and the European Union. The Quest for a Security/Defense Identity, 1948-1999*, in: Victor S. Papacosma, Sean Kay, Mark R. Rubin (eds.), NATO after Fifty Years, Wilmington 2001, pp. 141-161.

Leighton K. Marian: *Greco-Turkish Friction: Changing Balance in the Eastern Mediterranean*, Conflict Studies, No. 109, London 1979.

Levy S. Jack: *Domestic Politics and War*, in: Robert I. Rotberg, Theodore K. Rabb (eds.), The Origins and Prevention of Major Wars, Cambridge 1988, pp. 79-93.

Lindley-French Julian: *NATO, Britain, and the Emergence of a European Defense Capability*, in: Victor S. Papacosma, Sean Kay, Mark R. Rubin (eds.), NATO after Fifty Years, Wilmington 2001, pp. 43 – 51.

Lipson Charles: *International Cooperation in Economic and Security Affairs*, in: Baldwin A. David (ed.), Neorealism and Neoliberalism: The Contemporary Debate, New York 1993, pp. 60-80.

Lord Robertson: *Die Tragödie als Chance. Die NATO nach dem 11. September*, in: Internationale Politik, 7/2002, pp. 1-6.

Loukas Daphne: *Griechenland und die Türkei: Die ewigen Kontrahenten?*, Masters Thesis, University of Vienna, 1999.

Loulis C. John: *Papandreou`s Foreign Policy*, in: Foreign Affairs, Vol. 63, No. 2, Winter 1984/85, pp.375-391.

Mackenzie Kenneth: *Greece and Turkey: Disarray on NATO`s Southern Flank*, Conflict Studies, No. 154, London 1983.

Mango Adrew: *Turkish Policy in the Middle East. Turning Danger to Profit*, in: Clement H. Dodd (ed.), Turkish Foreign Policy: New Prospects, Occasional Papers, No.2, Huntingdon 1992, pp. 58-64.

Martin Lisa: *Interests, Power, and Multilateralism*, in: International Organization, Vol. 46, No. 4, Autumn 1992, pp. 765-792.

Martin L. Lisa & Beth A. Simmons: *Theories and Empirical Studies of International Institutuions*, in: International Organization, Vol. 52, No. 4, Autumn 1998, pp. 729-757.

Mastanduno Michael: *Do Relative Gains Matter? America`s Response to Japanese Industrial Policy*, in: Baldwin A. David (ed.), Neorealism and Neoliberalism: The Contemporary Debate, New York 1993, pp. 250-264.

Mavratsas V. Caesar: *Greek Cypriot Identity and Conflicting Interpretations of the Cyprus Problem*, in: Dimitris Keridis, Dimitirios Triantaphyllou (eds.), Greek – Turkish Relations in the Era of Globalization, Everett 2001, pp. 151-179.

McCalla B. Robert: *NATO`s Persistence after the Cold War*, in: International Organization, Vol. 50, Nr. 3, Summer 1996, pp. 445-475.

McDonald Robert: *Alliance Problems in the Eastern Mediterranean – Greece, Turkey and Cyprus: Part II*, in: IISS, Prospects for Security in the Mediterranean, Adelphi Papers, No. 229, London 1988, pp. 72-89.

McDonald Robert: *Greek – Turkish Relations and the Cyprus Conflict*, in: Dimitris Keridis, Dimitirios Triantaphyllou (eds.), Greek – Turkish Relations in the Era of Globalization, Everett 2001, pp. 116-150.

158

Mearsheimer J. John: *The False Promise of International Institutions*, in: International Security, Vol. 19, No. 3, Winter 1994/95, pp. 5-49.

Meinardus Roland: *Die Türkei-Politik Griechenlands – Der Zypern-, Ägäis-, und Minderheitenkonflikt aus der Sicht Athens (1967-1982)*, Frankfurt/Main 1982a.

Meinardus Roland: *Griechenlands gestörtes Verhältnis zur NATO*, in: Europa-Archiv, 4/1982b, pp. 105-114.

Meinardus Roland: *Der griechisch-türkische Konflikt über den Status der Ostägäischen Inseln*, in: Europa-Archiv, 2/1985, pp. 41-48.

Meinardus Roland: *Third-party Involvement in Greek-Turkish Disputes*, in: Dimitri Constas (ed.), The Greek – Turkish Conflict in the 1990s. Domestic and External Influences, London 1991, pp. 157-163.

De Mesquita B. Bruce: *The Contribution of Expected Utility Theory to the Study of International Conflict*, in: Robert I. Rotberg, Theodore K. Rabb (eds.), The Origins and Prevention of Major Wars, Cambridge 1988, pp. 53 – 73.

Michta A. Andrew: *Civil-Military Relations in the New NATO: The Standard and The Boundaries of Professionalism*, in: Victor S. Papacosma, Sean Kay, Mark R. Rubin (eds.), NATO after Fifty Years, Wilmington 2001, pp. 103-120.

Milner Helen: *International Theories of Cooperation among Nations. Strength and Weaknesses*, in: World Politics, Vol. 44, No. 3, April 1992, pp. 466-496.

Milner Helen: *The Assumption of Anarchy in International Relations Theory: A Critique*, in: Baldwin A. David (ed.), Neorealism and Neoliberalism: The Contemporary Debate, New York 1993, pp. 143-167.

Mosser W. Michael: *Engineering Influence: The Subtile Power of Small States in the CSCE/OSCE*, in: Erich Reiter, Heinz Gärtner (eds.), Small Statesand Alliances, Heidelberg 2001, pp. 63-84.

Nachmani Amikam: *What says the Neighbor to the West? On Turkish - Greek Relations*, in: Barry Rubin, Kemal Kirişci (eds.), Turkey in World Politics. An Emerging Multiregional Power, Boulder 2001, pp.71-89.

NATO: *NATO Handbuch. Jubiläumsausgabe zum fünfzigjährigen Bestehen der Nordatlantikpakt-Organisation*, Brüssel 1998.

Naumann Klaus: *Das Bündnis vor dem Aus? Gedanken über die Zukunft der NATO*, in: Internationale Politik, 7/2002, pp. 7-14.

Nicolaidis Kalypso: *Europe's Tainted Mirror. Reflections on Turkey's Candidacy Status after Helsinki*, in: Dimitris Keridis, Dimitirios Triantaphyllou (eds.), Greek – Turkish Relations in the Era of Globalization, Everett 2001, pp. 245-278.

Nicolet Claude: *American and British NATO-Plans for Cyprus, 1959-1964*, in: Thetis, 8/2001, pp. 314-318.

Nye S. Joseph (Jr.): *Old Wars and Future Wars: Causation and Prevention*, in: Robert I. Rotberg, Theodore K. Rabb (eds.), The Origins and Prevention of Major Wars, Cambridge 1988, pp. 3 –34.

Nye S. Joseph & Robert O. Keohane: *The United States and International Institutions in Europe after the Cold War*, in: Robert O. Keohane et al. (eds.), After the Cold War: International Institutions and State Strategies in Europe, 1989-1991, Cambridge 1993.

Oran Baskin: *Türk Dış Politikasi ve Batı Trakya*, in: Faruk Sönmezoğlu (ed.),Türk Dış Politikasının Analizi, İstanbul 2001, pp.307-320.

Orhun Ömür: *European Security and Defence Identity – Common European Security and Defence Policy: A Turkish Perspective*, in: Perceptions - Journal of International Affairs, Vol. 5, No. 3, September-November 2000, 115-123.

Osgood E. Robert: *The Nature of Alliances*, in: Robert O. Matthews, Arthur G. Rubinoff, Janice Gross Stein (eds.), International Conflict and Conflict Management, Ontario 1989, pp.458-462.

Özcan Gencer: *Türkiye`de Siyasal Rejim ve Dış Politika: 1983-1993*, in: Faruk Sönmezoğlu (ed.), Türk Dış Politikasının Analizi, İstanbul 2001a, pp. 511-534.

Özcan Gencer: *The Military and the Making of Foreign Policy in Turkey*, in: Barry Rubin, Kemal Kirişci (eds.), Turkey in World Politics. An Emerging Multiregional Power, Boulder 2001b, pp.13-30.

Pabst Martin: *Neue Bemühungen zur Beilegung des Zypernkonfliktes*, in: ÖMZ, 5/2000, pp. 567-574.

Papacosma S. Victor: *NATO and the Balkans*, in: Victor S. Papacosma, Mary Ann Heiss (eds.), NATO in the Post-Cold War Era: Does it have a Future?, Houndsmill 1993, pp. 247 – 271.

Papacosma S. Victor: *NATO and Internal Disputes. Greece and Turkey*, in: Victor S. Papacosma, Sean Kay, Mark R. Rubin (eds.), NATO after Fifty Years, Wilmington 2001, pp. 199-225.

Papacosma S. Victor, Sean Kay & Mark R. Rubin (eds.): *NATO after Fifty Years*, Wilmington 2001.

Papahadjopoulos Daphne: *Greek Foreign Policy in the Post-Cold War Era: Implications for the European Union*, CEPS Paper, No. 2, Brussels 1998.

Papandreou A. Georgios: *Greece and Its Neighbours in 21st Century Europe*, in: Südosteuropa Mitteilungen, Vol. 39, No. 1, 1999, pp. 1-8.

Peters Ingo: *The OSCE and German Policy: A Study in How Institutions Matter*, in: Helga Haftendorn, Robert O. Keohane, Celeste A. Wallander (eds.), Imperfect Unions – Security Institutions over Time and Space, Oxford 1999, pp. 195-220.

160

Pevehouse C. Jon: *Democracy from the Outside-In? International Organizations and Democratization*, in: International Organization, Vol. 56, No. 3, Summer 2002, pp. 515-549.

Platias Athanasios: *Greece`s Strategic Doctrine: In Search of Autonomy and Deterrence*, in: Dimitri Constas (ed.), The Greek – Turkish Conflict in the 1990s. Domestic and External Influences, London 1991, pp. 91-108.

Pleninger Johann: *Die Sicherheitspolitische Bedeutung der Türkei für die NATO*, Masters Thesis, University of Vienna, 2002.

Polyviou G. Polyvios: *Conflict and Negotiation: 1960 – 1980*, New York 1980.

Pourchot V. Georgeta: *NATO Enlargement and Democracy in Eastern Europe*, in: European Security, Vol. 6, No. 4, Winter 1997, pp. 157-174.

Powell Robert: *Absolute and Relative Gains in International Relations Theory*, in: Baldwin A. David (ed.), Neorealism and Neoliberalism: The Contemporary Debate, New York 1993, pp. 209-230.

Pridham Geoffrey: *Linkage Politics Theory and the Greek – Turkish Reapprochment*, in: Dimitri Constas (ed.), The Greek – Turkish Conflict in the 1990s. Domestic and External Influences, London 1991, pp. 73-88.

Pridham Geoffrey: *The Dynamics of Democratization – A Comparative Approach*, London 2000.

Rearden L. Steven: *NATO's Strategy: Past, Present, and Future*, in: Victor S. Papacosma, Mary Ann Heiss (eds.), NATO in the Post-Cold War Era: Does it have a Future?, Houndsmill 1993, pp. 70 – 89.

Rearden L. Steven: *NATO`s Post-Cold War Strategy. The Role of Combined Joint Task Forces*, in: Victor S. Papacosma, Sean Kay, Mark R. Rubin (eds.), NATO after Fifty Years, Wilmington 2001, pp. 75-86.

Reiter Dan: *Why NATO Enlargement does not Spread Democracy*, in: International Security,Vol. 25, No. 4, Spring 2001, pp. 41-67.

Reiter Erich & Heinz Gärtner (eds.): *Small States and Alliances*, Heidelberg 2001.

Reuter Jürgen: *Athens Türkeipolitik im Wandel – Griechisch-türkische Beziehungen vor und nach dem EU-Gipfel von Helsinki*, in: Südosteuropa Mitteilungen, Vol. 40, Nr. 1, 2000, pp. 47-64.

Riecke Henning: *Nukleare Nichtverbreitung als Aktionsfeld von NATO und GASP*, in: Helga Haftendorn, Otto Keck (eds.), Kooperation jenseits von Hegemonie und Bedrohung. Sicherheitsinstitutionen in den internationalen Beziehungen, Baden-Baden 1997, pp.191-232..

Riedel Sabine: *Die griechisch-türkischen Spannungen vor dem Hintergrund des Kriegs im ehemaligen Jugoslawien. Neue Aspekte eines alten Konfliktherds in Südosteuropa*, in:Südosteuropa, Vol. 45, No. 1, 1996, pp. 11-47.

Riemer K. Andrea: *Die Armee in der Türkei: Chancen- oder Risikopotential?*, in: Südosteuropa, Vol. 48, No, 9-10, 1999. pp. 521-557.

Riemer K. Andrea: *Griechenland und Türkei im neuen Millenium – Stabilisierer versus Regionalmacht*, Frankfurt/Main 2000.

Riemer K. Andrea & Yannis A. Stivachtis: *Turkey and Greece: Quo Vadis?*, in: ÖMZ, 5/2000, pp. 559-567.

Robins Philip: *Turkish Policy and the Gulf Crisis: Adventurous or Dynamic?*, in: Clement H. Dodd (ed.), Turkish Foreign Policy: New Prospects, Occasional Papers, No. 2, Huntingdon 1992, pp. 70-86.

Roper John: *The West and Turkey: Varying Roles, Common Interest*, in: The International Spectator, Vol. 34, No. 1, January-March 1999a, pp. 89-102.

Roper John: *NATO's New Role in Crisis Management*, in: The International Spectator, Vol. 34, No. 2, April – June 1999b, pp. 51-61.

Roubatis P. Yiannis: *The United States and the Operational Responsibilities of the Greek Armed Forces, 1947-1987*, in: Journal of Hellenic Diaspora, Vol. 6, No. 1, Spring 1979, pp. 39-57.

Roubatis P. Yiannis: *Tangled Webs - The US in Greece 1947-1967*, New York 1987.

Rubin Barry: *Turkey: A Transformed International Role*, in: Barry Rubin, Kemal Kirişci (eds.), Turkey in World Politics. An Emerging Multiregional Power, Boulder 2001a, pp. 1-4.

Rubin Barry: *Understanding Turkey's New Foreign Policy*, in: Barry Rubin, Kemal Kirişci (eds.), Turkey in World Politics. An Emerging Multiregional Power, Boulder 2001b, pp. 251-254.

Rühl Lothar: *NATO's Political Limitations*, in: The Atlantic Community Quarterly, Vol. 12, No. 4, Winter 1974-75, pp. 463-480:

Rühl Lothar: *Die Zypernkrise von 1974 und der griechisch-türkische Interessenkonflikt*, in: Europa-Archiv, 22/1975, pp. 699-710.

Rühl Lothar: *Der Zypernkonflikt, die Weltmächte und die Europäische Sicherheit*, in: Europa-Archiv, 1/1976, pp. 19-30.

Rupp W. Rainer: *Burden Sharing and the Southern Region of the Atlantic Alliance*, in: Stuart T. Douglas (ed.), Politics and Security in the Southern Region of the Atlantic Alliance, Houndsmill 1988, pp. 27-45.

Russett Bruce, John R. Oneal & David R. Davis: *The Third Leg of the Kantian Tripod for Peace: International Organizations and Militarized Disputes, 1950-85*, in: International Organization, Vol. 52, No. 3, Summer 1998, pp. 441-467.

Sasley Brent: *Turkey's Energy Politics*, in: Barry Rubin (ed.),Turkey in World Politics. An Emerging Multiregional Power, Boulder 2001, pp. 217-230.

Schake Kori: *Arms Control after the Cold War. The Challenge of Diverging Security Agendas*, in: Victor S. Papacosma, Sean Kay, Mark R. Rubin (eds.), NATO after Fifty Years, Wilmington 2001, pp. 191-198.

Schmidl A. Erwin: *Small States and International Operations*, in: Erich Reiter, Heinz Gärtner (eds.), Small States and Alliances, Heidelberg 2001, pp. 85-88.

Schweller L. Randall: *Bandwagoning for Profit – Bringing the Revisionist State Back In*, in: International Security, Vol. 19, No. 1, Summer 1994, pp. 72-107.

Sebenius K. James: *Negotiation Arithmetic: Adding and Subtracting Issues and Parties*, in: International Organization, Vol. 37, No. 2, Spring 1983, pp. 281-316.

Segal Gerald: *International Relations and Democratic Transition*, in: Geoffrey Pridham (ed.), Encouraging Democracy – The International Context of Regime Transition in Southern Europe, Leicester 1991, pp. 31-44.

Şenesen Günlük Gülay: *Türk Silahli Kuvvetlerinin Modernizasyon Programının Bir Değerlendirmesi*, in: Faruk Sönmezoğlu (ed.), Türk Dış Politikasının Analizi, İstanbul 2001, pp. 585-604.

Sens G. Allen: *From Collective Defense to Cooperative Security? The New NATO and Nontraditional Challenges and Mission*, in: Victor S. Papacosma, Sean Kay, Mark R. Rubin (eds.), NATO after Fifty Years, Wilmington 2001, pp. 165-189.

Serfaty Simon: *The Management of Discord in Alliance Relations*, in: Stuart T. Douglas (ed.), Politics and Security in the Southern Region of the Atlantic Alliance, Houndsmill 1988, pp. 7-26.

Sezer B. Duygu: *Turkey's Security Policies*, Adelphi Papers, No. 164, London 1981.

Sezer Duygu: *The Strategic Matrix of the SEM: A Turkish Perspective*, in: Dimitri Constas (ed.), The Greek – Turkish Conflict in the 1990s. Domestic and External Influences, London 1991, pp. 109-126.

Sezer Bazoğlu Duygu: *Russia: The Challenges of Reconciling Geopolitical Competition with Economic Partnership*, in: Barry Rubin (ed.), Turkey in World Politics. An Emerging Multiregional Power, Boulder 2001, pp. 152-166.

Simon Jeffrey: *NATO Enlargement. Crossing the Rubicon*, in: Victor S. Papacosma, Sean Kay, Mark R. Rubin (eds.), NATO after Fifty Years, Wilmington 2001, pp. 121-139.

Skalnes S. Lars: *From the Outside In, From the Inside Out: NATO Expansion and International Relations Theory*, in: Security Studies, Vol. 7, No. 4, Summer 1998, pp. 44-87.

Sloan R. Stanley: *Continuity or Change? The View from America*, in: Victor S. Papacosma, Sean Kay, Mark R. Rubin (eds.), NATO after Fifty Years, Wilmington 2001, pp. 3 – 24.

163

Smith A. Martin: *At Arm's Length: NATO and the United Nations in the Cold War Era*, in: International Peacekeeping, Vol. 2, No. 1, Spring 1995, pp. 56-73.

Snidal Duncan: *Relative Gains and the Pattern of International Cooperation*, in: Baldwin A. David (ed.), Neorealism and Neoliberalism: The Contemporary Debate, New York 1993, pp. 170-201.

Snyder H. Glenn: *The Security Dilemma in Alliance Politics*, in: Journal of International Affairs, Vol. 36, No. 4, July 1984, pp. 461-495.

Sönmezoğlu Faruk: *ABD`nin Kıbrıs Politikası (1964-1980)*, İstanbul 1995.

Spaak Paul-Henri: *Memoiren eines Europäers*, Hamburg 1969.

Stearns Monteagle: *Entangled Allies – U.S. Policy toward Greece, Turkey, and Cyprus*, New York 1992.

Stearns Monteagle: *The Security Domain. A US Perspective*, in: Dimitris Keridis, Dimitirios Triantaphyllou (eds.), Greek – Turkish Relations in the Era of Globalization, Everett 2001, pp. 239-244.

Stein A. Arthur: *The Politics of Linkage*, in: World Politics, Vol. 33, No. 1, October 1980, pp. 62-81.

Stein Arthur: *Coordination and Collaboration Regimes in an Anarchic World*, in: Baldwin A. David (ed.), Neorealism and Neoliberalism: The Contemporary Debate, New York 1993, pp. 29-59.

Steinbach Udo: *Die Türkei – Regionale Macht oder Verbündeter des Westens?*, in: Erich Reiter (ed.), Jahrbuch für internationale Sicherheitspolitik 1999, Hamburg 1999, pp. 562-578.

Stephanou Constantine & Charalambos Tsardanides: *The EC Factor in the Greece – Turkey – Cyprus Triangle*, in: Dimitri Constas (ed.), The Greek – Turkish Conflict in the 1990s. Domestic and External Influences, London 1991, pp. 207-230.

Stikker Dirk: *The Role of the Secretary General of NATO*, in: Edgar S. Furniss, Jr. (ed.), The Western Alliance. Its Status and Prospects, Columbus 1965, pp. 3-28.

Stikker U. Dirk: *Bausteine für eine neue Welt. Gedanken und Erinnerungen an schicksalhafte Nachkriegsjahre*, Wien 1966.

Stivachtis A. Yannis: *Living with Dilemmas: Greek-Turkish Relations at the Rise of the 21st Century*, in: Erich Reiter (ed.), Jahrbuch für Internationale Sicherheitspolitik 2000, Hamburg 2000, pp. 473-494.

Stivachtis A. Yannis: *Co-Operative Security and Non-Offensive Defence in the Zone of War. The Greek-Turkish and the Arab-Israeli Cases*, Frankfurt am Main 2001.

164

Tams Carsten: *The Functions of a European Security and Defence Identity and Its Institutional Form*, in: Helga Haftendorn, Robert O. Keohane, Celeste A. Wallander (eds.), Imperfect Unions – Security Institutions over Time and Space, Oxford 1999, pp. 80-103.

Taylor Phillip: *Weapons Standardization in NATO: Collaborative Security or Economic Competition?*, in: International Organization, Vol. 36, No. 1, Winter 1982, pp. 95-112.

Theiler Olaf: *Der Wandel der NATO nach dem Ende des Ost-West-Konfliktes*, in: Helga Haftendorn, Otto Keck (eds.), Kooperation jenseits von Hegemonie und Bedrohung. Sicherheitsinstitutionen in den internationalen Beziehungen, Baden-Baden 1997, pp.101-136.

Theophanous Andreas: *The Cyprus Problem and its Implications for Stability and Security in the Eastern Mediterranean*, in: Dimitris Keridis, Dimitirios Triantaphyllou (eds.), Greek – Turkish Relations in the Era of Globalization, Everett 2001, pp. 180-207.

Tovias Alfred: *US Policy towards Democratic Transition in Southern Europe*, in: Geoffrey Pridham (ed.), Encouraging Democracy – The International Context of Regime Transition in Southern Europe, Leicester 1991, pp. 175-194.

Triantaphyllou Dimitrios: *Further Turmoil Ahead?*, in: Dimitris Keridis & Dimitirios Triantaphyllou (eds.), Greek – Turkish Relations in the Era of Globalization, Everett 2001, pp. 56-79.

Tülümen Turgut: *Hayat Boyu Kıbrıs*, İstanbul 1998.

Turan İlter: *Political Parties and the Party System in post-1983 Turkey*, in: Metin Heper, Ahmet Evin (eds.), State, Democracy and the Military. Turkey in the 1980s, Berlin 1988, pp.63-80.

Turan İlter & Dilek Barlas: *Bati Ittifakına Üye Olmanın Türk Dış Politikası Üzerindeki Etkileri*, in: Faruk Sönmezoğlu (ed.), Türk Dış Politikasının Analizi, İstanbul 2001, pp. 647-662.

Tuschhoff Christian: *Alliance Cohesion and Peaceful Change in NATO*, in: Helga Haftendorn, Robert O. Keohane, Celeste A. Wallander (eds.), Imperfect Unions – Security Institutions over Time and Space, Oxford 1999, pp. 140-161.

Tuschoff Christian: *Gaining Control*, in: Erich Reiter, Heinz Gärtner (eds.), Small States and Alliances, Heidelberg 2001, pp. 51-61.

Tzermias Pavlos: *Polarisierung in Griechenland. Knapper PASOK-Sieg bei den Parlamentswahlen*, in: Europäische Rundschau, Vol. 28, No. 2, 2000, pp 85-92.

Ülman A. Haluk: *Türk Dış Politikasına Yön Veren Etkenler (1923-1968) I*, in: Atatürk Üniversitesi Siyasal Bilgiler Fakültesi Dergisi, Vol. 23, No. 3, 1968, pp. 241-273.

Ülman A. Haluk & Oral Sander: *Türk Dış Politikasına Yön Veren Etkenler (1923-1968) II*, in: Atatürk Üniversitesi Siyasal Bilgiler Fakültesi Dergisi, Vol. 27, No. 1, 1972, pp. 1-25.

Ünal Hasan & Demirtaş-Coşkun Birgül: *Kıbrıs Meselesi: Kisa Tarihçe, Mevcut Durum Analizi ve Muhtemel Senaryolar*, in: Stratejik Analiz, Vol. 1, No. 12, April 2001, pp. 40-51.

Uslu Nasuh: *Türk Amerikan İlişkilerinde Kıbrıs*, Ankara 2000.

Uzgel Ilhan: *The Balkans: Turkey's Stabilizing Role*, in: Barry Rubin (ed.), Turkey in World Politics. An Emerging Multiregional Power, Boulder 2001a, pp. 49-66.

Uzgel İlhan: *ABD ve NATO'yla İlişkiler* [1980-1990], in: Baskın Oran (ed.), Türk Dış Politikası. Kurtuluş Savaşından Bugüne Olgular, Belgeler, Yorumlar,Volume II, Istanbul 2001b, pp. 34-81.

Uzgel İlhan: *ABD ve NATO'yla İlişkiler* [1990-2001], in: Baskın Oran (ed.), Türk Dış Politikası. Kurtuluş Savaşından Bugüne Olgular, Belgeler, Yorumlar, Volume II, Istanbul 2001c, pp. 243-325.

Uzgel İlhan: *Balkanlar'la İlişkiler* [1990-2001], in: Baskın Oran (ed.), Türk Dış Politikası. Kurtuluş Savaşından Bugüne Olgular, Belgeler, Yorumlar, Volume II, Istanbul 2001d, pp. 519-523.

Vamik D. Volkan & Norman Itzkowitz: *Turks and Greeks: Neighbours in Conflict*, Huntingdon 1994.

Varvaroussis Paris: *Konstellationsanalyse der Außenpolitik Griechenlands und der Türkei (1974-75: seit der Invasion der Türkei in Zypern)*, München 1979.

Varvaroussis Paris: *Neue machtpolitische Konstellation zwischen Griechenland und der Türkei*, in: Südosteuropa, Vol. 42, No. 9, 1993, pp. 534-549.

Varvaroussis Paris: *Die griechische Diplomatie auf dem Balkan – Rückkehr in die Geschichte?*, in: Südosteuropa, Vol. 44, No. 6-7, 1995, 373-384.

Varwick Johannes & Wicahrd Woyke: *NATO 2000 – Atlantische Sicherheit im Wandel*, Opladen 1999.

Veremis Thanos: *Greek Security Issues and Politics*, in: Alford Jonathan (ed.), Turkey and Greece: Adversity in Alliance, Aldershot 1984, pp. 1-38.

Veremis Thanos: *Greece*, in: Stuart T. Douglas (ed.), Politics and Security in the Southern Region of the Atlantic Alliance, Houndsmill 1988a, pp. 137-156.

Veremis Thanos: *Greece and NATO: Continuity and Change*, in: John Chipman (ed.), NATO's Southern Allies: Internal and External Challenges, London 1988b, pp. 236-286.

Veremis Thanos: *Defence and Security Policies under PASOK*, in: Richard Clogg (ed.), Greece 1981-1989: The Populist Decade, London 1993.

Veremis Thanos: *The Protracted Crisis*, in: Dimitris Keridis, Dimitirios Triantaphyllou (eds.), Greek – Turkish Relations in the Era of Globalization, Everett 2001, pp. 42-55.

Verney Susannah & Theodore Couloumbis: *State – International Systems Interaction and the Greek Transition to Democracy in the mid-1970s*, in: Geoffrey Pridham (ed.), Encouraging Democracy – The International Context of Regime Transition in Southern Europe, Leicester 1991, pp. 103-124.

Wallander A. Celeste: *Institutional Assests and Adaptability: NATO after the Cold War*, in: International Organization, Vol. 54, No. 4, Autumn 2000, pp. 705-732.

Wallander A. Celeste & Robert O. Keohane: *Risk, Threat, and Security Institutions*, in: Helga Haftendorn, Robert O. Keohane, Celeste A. Wallander (eds.), Imperfect Unions – Security Institutions over Time and Space, Oxford 1999, pp. 21-47.

Waltz Kenneth: *Theory of International Politics*, New York 1979.

Waltz Kenneth: *The Origins of War in Neorealist Theory*, in: Robert I. Rotberg, Theodore K. Rabb (eds.), The Origins and Prevention of Major Wars, Cambridge 1988, pp. 39-51.

Waltz Kenneth: *The Emerging Structure of International Politics*, in: International Security, Vol. 18, No. 2, Fall 1993, pp. 44-79.

Weber Steve: *Shaping the Postwar Balance of Power: Multilateralism in NATO*, in: International Organization, Vol. 46, No. 3, Summer 1992, pp. 633-680.

Weitsman A. Patricia: *Intimate Enemies: The Politics of Peacetime Alliances*, in: Security Studies, Vol. 7, No. 1, Autumn 1997, pp. 156-192.

Wilson Andrew: *The Aegean Dispute*, Adelphi Papers, No. 155, London 1979.

Windsor Philip: *NATO and the Cyprus Crisis*, Adelphi Papers, No. 14, London 1964.

Winrow Gareth: *Turkey and the Newly Independent States of Central Asia and the Transcaucasus*, in: Barry Rubin & Kemal Kirişci (eds.), Turkey in World Politics. An Emerging Multiregional Power, Boulder 2001, pp. 173-185.

Wohlstetter Albert: *Die Türkei und die Sicherung der Interessen der NATO*, in: Europa-Archiv, Vol. 40, No. 16, 1985, pp. 507-514.

Woodhouse M. C.: *Karamanlis: The Restorer of Greek Democracy*, Oxford 1982.

Yost S. David: *The New NATO and Collective Security*, in: Survival, Vol. 40, No. 2, Summer 1998, pp. 135-160.

Newspapers (mainly web editions) and News Agencies:

Athens News Agency (ANA), www.hri.org/news/greek/ana
Hürriyet, www.hurriyet.com.tr
International Herald Tribune (IHT), www.iht.com
Kathimerini (English edition), www.ekathimerini.com
Neue Zürcher Zeitung (NZZ), www.nzz.ch
The New York Times (NYT), www.nyt.com
ORF.at, www.orf.at
Die Presse, www.diepresse.at
Radikal, www.radikal.com.tr
Turkish Daily News (TDN), www.turkishdailynews.com
The Washington Post (WP), www.washingtonpost.com

INTERNATIONALE SICHERHEIT

Herausgegeben von Heinz Gärtner

Band 1 Thomas Pankratz: Möglichkeiten und Grenzen einer europäischen Rüstungskooperation. Dargestellt an der Theorie der Relative Gains. 2002.

Band 2 Karl Schmidseder: Internationale Interventionen und *Crisis Response Operations*. Charakteristika, Bedingungen und Konsequenzen für das internationale und nationale Krisenmanagement. 2003.

Band 3 Reiner Meyer: Ein Friedensprozeß ohne Versöhnung. Der Israelisch-Palästinensische Konflikt und die Oslo-Verhandlungen als Beispiel für die Probleme des Konfliktmanagements. 2004.

Band 4 Ewelina Hilger: Präemption und humanitäre Intervention – gerechte Kriege? 2005.

Band 5 Hakan Akbulut: NATO's Feuding Members: The Cases of Greece and Turkey. 2005.

www.peterlang.de